ARCANA OF NATURE;

OR THE

PHILOSOPHY OF SPIRITUAL EXISTENCE

AND OF

THE SPIRIT WORLD.

BY

HUDSON TUTTLE.

HEAVEN, THE HOME OF THE IMMORTAL SPIRIT, IS ORIGINATED AND SUSTAINED BY NATURAL LAWS.

VOL. II.

BOSTON:
WILLIAM WHITE & CO.,
158 WASHINGTON STREET.
1864.

Entered, according to Act of Congress, in the year 1860, by
HUDSON TUTTLE,
In the Clerk's Office of the District Court of the District of Massachusetts.

STEREOTYPED AT THE
BOSTON STEREOTYPE FOUNDRY.

TO THE

FREE, UNTRAMMELLED THINKERS

OF

EVERY NATION, COUNTRY, AND RACE,

This Volume

IS RESPECTFULLY DEDICATED

BY THE AUTHOR.

PREFACE.

I SALUTE the public in a second volume, extending the philosophy of the first into a deeper and more mysterious domain. I can make no apology to the reader, nor extenuation whatever. If he is profited by the perusal of these pages, let him praise the real authors, and not the labors of one who, like himself, is taught by them; if he blames and is disgusted, not on me, but on the invisible authors, the censure falls. In my humble sphere I have toiled, not to introduce my own thoughts,* but to receive those of higher intelligences as purely and truthfully as possible, and transmit them to the consideration of those who may be attracted to my pages. That such attraction may prove a source of profit, noble and true thoughts be inculcated, and that this volume may not be without influence for goodness and truth, is my earnest prayer.

H. T.

WALNUT GROVE FARM.

* I have introduced a few passages, but have carefully included them in brackets, as I have no desire to confound their authorship with that of the volume.

CONTENTS.

CHAPTER I.

EVIDENCES OF MAN'S IMMORTALITY DRAWN FROM HISTORY. — SPIRITUALISM OF THE NATIONS.

CHAPTER II.

PROOFS OF IMMORTALITY DRAWN FROM HISTORY, CONCLUDED.

CHAPTER III.

EVIDENCES OF MAN'S IMMORTALITY DERIVED FROM MODERN SPIRITUALISM.

CHAPTER IV.

THE OBJECTS OF MODERN SPIRITUALISM.

CHAPTER V.

CONSIDERATION OF SPIRITUAL PHENOMENA, AND THEIR DISTINCTION FROM SUCH AS ARE NOT SPIRITUAL, BUT DEPENDENT ON SIMILAR LAWS.

CHAPTER VI.

SPACE ETHER.

CHAPTER VII.

PHILOSOPHY OF THE IMPONDERABLE AGENTS IN THEIR RELATION TO SPIRIT.

CHAPTER VIII.

PHILOSOPHY OF THE IMPONDERABLE AGENTS IN THEIR RELATIONS TO SPIRIT, CONCLUDED.

CHAPTER IX.

THE IMPONDERABLE AGENTS AS MANIFESTED IN LIVING BEINGS.

CHAPTER X.

SPIRITUAL ELEMENTS.

CHAPTER XIV.

PHILOSOPHY OF CHANGE AND DEATH, CONCLUDED.

CHAPTER XV.

SPIRIT, ITS ORIGIN, FACULTIES, AND POWER.

CHAPTER XVI.

A CLAIRVOYANT'S VIEW OF THE SPIRIT SPHERE.

CHAPTER XVII.

PHILOSOPHY OF THE SPIRIT WORLD.

CHAPTER XVIII.

SPIRIT LIFE.

THE

ARCANA OF NATURE.

INTRODUCTION.

FROM the realm of physical causes, the so-called World of Matter, we have arisen, step by step, to the domain of Spirit. We pass from the world of the senses to the mysterious, invisible world beyond their ken. We pass the line of demarcation between Matter and Spirit. But no reasoning can be sound if based on hypothetical data. If we conceive of spirit at all, it must be through the medium of matter. If not composed of matter, it is nought. Reason cannot refute the oft-repeated maxim: "Something cannot originate from nothing; an infinitude of nothings is nothing still."

This view may be considered materialistic. It is truly so, but it is not materialism as commonly understood. Philosophers investigate matter, its attributes and laws, as far as the limited range of their senses; and when it becomes too attenuated to reveal itself, they complacently call it spirit, and assign for it a confused and contradictory existence.

The line of demarcation drawn by them may be preserved, as it is a convenient designation; yet it is false in theory and fact.

All things which exist are material; without matter nothing exists.

This is the sublime axiom on which the present volume is based. We know of nothing which conflicts with it, except the theories of men, some of which are entitled to respect, but are incapable of bearing the test of reason.

As the human mind is an epitome of Nature, as it is constructed to understand and investigate surrounding creation, we must look to its mysterious workings for the true explanation of the phenomena we fail elsewhere to understand. We look on its failure to comprehend the creation of something from nothing as a significant fact pointing to a truthful solution of the question —

What is spirit? Is it a fog, a vapor? Who can comprehend, who define it? Man dies. His spirit is immortal; so we are taught. How does it exist? Has it an identified being, or is it, like Nature in the Indian cosmogony, absorbed, after a fleeting existence, into the bosom of the infinite Bram, the great fountain of spirit, to flow out again in perpetual cycles of evanescent forms? What a horrid picture for the thinking man! We cry, "O, let our selfhood be preserved. If we exist, let us remain as we are." Such is the soul's aspiration. Any other state of being is non-existence. If consciousness is lost, all is lost; for eternal death and eternal sleep are one. The manner of the future existence is the problem — whither cometh the soul, whither goeth?

Clairvoyance has given us a clear response. It has led us deeper into the mysterious vale of spirit than any sage has ever done; and, except the recent developments of spiritualism, affords the only insight to be obtained of the inner life of men. It proves man to be composed of spirit as well as body, or, to use the words of another, who has forcibly expressed the truth known to the ancient sages, "Man is an intelligence served by organs."

When, as in the preceding volume, man is studied from the material stand-point, there seems no reason why he should be immortal. The mind appears to be an emanation from the elementary combinations in the physical body, and as the hum of the bee is no longer heard after the insect has passed, so mind dies when the form which calls it forth expires.

But here a new light dawns. We pour the bright beams of LAW, studied in the physical world, on the philosophy of spiritual existence.

In the physical realm we shall learn the origin of the spirit, and by questioning spirits learn the grandeur of man's immortal destiny.

Sublime beauties unfold to our enraptured vision here on the threshold of the unseen world. We stand enchanted by indescribable magnificence. Eternal progress is the law of spirit as well as of matter; and how speak of the spirit sage of a thousand ages?

We pause for a moment to review the dogmatic theories the world in its ignorance has entertained. Our task is light, for, in reality, it knows little of truth. What can the dark materialist know of spirit? What the so-called spiritualist who attempts to rob existence of matter? Stupendous systems of theology, time-hoary volumes of saints and fathers, we revere you. We recognize the attempts you have made, though they have been abortive. But your day has passed. The present demands a more satisfying system than a childish play on words, the polemics of the schools, the cant of the doctors.

Such this volume will strive to unfold — to prove the immortality of spirit, and the manner of its existence in the spirit world, its origin, law, and destiny.

We shall strive to retain the positive method of treating our subject as rigorously as in the former volume, though perhaps the reader who does not grant our position may not think we do so. But we have only to tell him that we receive clairvoyance as positive testimony, and assume spirit intercourse as an admitted fact.

CHAPTER I.

EVIDENCES OF MAN'S IMMORTALITY DRAWN FROM HISTORY. — SPIRITUALISM OF THE NATIONS.

Universality of the Belief in future Existence. — Teachings of Nature. — When was this Problem solved? — Records of the Hindoos. — Their sacred Books. — Of the Hebrews, Ascetics, Hermits, Power of Spirit, Persian and Chaldean Beliefs. — Ancient Sages. — Greeks. — Poets. — Hesiod. — Mythology. — The Middle Passage. — Epimenides of Crete. — Cassandra, Princess of Troy. — The Solution.

AMONG nearly all tribes and races of men exists a deep and abiding faith in immortal existence. It enters the heart by intuition, and there moulds a beautiful, ethereal creation, peopled by mythic dreams and wild fantasies, yet ever fostering the cardinal idea which the human mind in every stage of its development so loves — its own immortality. To the reflecting mind the universality of this belief becomes a strong proof of man's future existence. It is so contrary to the effects transpiring around us, so foreign to the course observed in objective nature, that it would almost seem the voice of a superior being must have whispered it to man in a gush of inspiration.

Observe the phenomena transpiring around us. Do we see any trace of immortality? any clew to a higher spiritual state after the death of the body? Rather of destruction. The tree which blooms to-day, the pride of the vegetable world, to-morrow lies in the dust, and in a few years nothing is left of its proud stem. It has gone into the atmosphere and the soil, to support other organisms. Generation after generation of forests have thus decayed. From the wreck of the old the new receives birth. Nature works in a mighty cycle, ever returning within itself. Nothing is lost. The old particles are absorbed, and it matters not whether the atoms of the mouldering oak are carried by the winds to nourish the palms waving their delicate foliage in the tropical breeze, or to sustain the physical system of man. Her end is answered. Nothing is lost. New forms spring from the old, and the perpetual circulation never rests.

To savage man, in his rude estate, this wonderful scheme of compensation is incomprehensible. Like children they view the turmoil of the elements, wholly ignorant of the causes underlying the effects which are presented to their awe-struck imagination. They must be impressed with this decay and destruction in all its frightful deformity, unmitigated by the compensation of renovation. They see their companions die, become inanimate masses of flesh, presenting the same symptoms when dying, and aspect after death, as the animal shot with their arrow. No circumstance indicates the future to their rude minds; yet, standing there beside the lifeless corses of their friends, they worked out the grandest problem that can be presented to the mind of man — his own eternal existence.

At first their solution assumed the crude forms of their own minds, for it is with difficulty the greatest philosopher can grasp the manner and form of the spiritual essence. Could the immediate growth of vegetation, from the decaying atoms of the old, first suggest the idea of the transmigration of the life principle from the dying plant to the germinating one?

from the animal gasping in the pangs of death to the one gasping its first breath of life? However this may be, the earliest legends, which are handed down from the mythic ages, represent the rude savage as already believing in a complicated system of mythology, which entertained the immortality of man, and the form and method of his future life — a ceaseless transmigration from one form of existence to another.

The problem was solved by the men who, as it were, stood on the verge of time. How did they arrive at its solution? Not through the senses, for these taught eternal death; not by reason, research, or reflection. There must have been a time in the remote ages when this belief was not entertained — when man was too savage to receive it. If so, at some definite period he must have been enlightened on this vitally important subject. How else could he have received it but by the voice of his own spirit? If the spirit is immortal, should it not know the destiny which awaits it? Reason answers yes. Then is the early shadowing forth of the future life readily explained. It is the yearning of the immortal spirit, conscious of its godlike destiny, striving to embody its intuition in words.

The earliest authentic records of mankind are accorded to the Hindoos. So soon had a complicated system of mythology been worked out of the fundamental idea of immortality. Their sacred books teach that Bram is the Eternal Spirit, from which all existence flows, and back to which all returns. He causes a mighty ebb and flow of creation, death, and renovation, in a perpetual return, perfectly compensating itself. From him all grades of intelligences came, from those scarcely his inferiors, to man. They believe that every man is accompanied through life by two spirits; one keeps an account of his good, the other of his evil deeds. They believe that within the external, mortal body resides a spiritual body, from which the mind emanates — a true conception, and more wonderful for being so early learned. After various probations on earth, in hell, and paradise, the spirit casts off this spiritual form, becomes wholly freed from matter, and is completely absorbed in Bram, the great fountain from which it came. The spirit-body returns to be again born on earth. These ideas are vague, yet they contain a great truth — the eternal progress of spirit, in refinement and elevation, until, too sublimated to be comprehended by man, it becomes lost in infinity, as the eagle soaring upward vanishes in the empyrean.

Man is acted on by a host of invisible intelligences, some of which influence his passions, others drown him with the lethargy of ignorance; and only by the most determined efforts can he cast their detrimental influence aside.

Their sacred books describe fourteen spheres, the abode of spirits. The earth is one of these spheres, having six gradations of paradise above, and seven of punishment below it, each more terrible than the other. These spheres are described as being formed of red hot copper, thickly set with thorns, or festering with deadly serpents, while the lowest is a pit heated to redness with burning charcoal. When a man dies, his soul is immediately conducted to Gama, the judge of the dead, before whom is laid the record of his life. Then, if he is sinful, and has led a wicked life, he is given to evil spirits, who come up from the lower spheres, and they, according to his sentence, drag him over rocky paths, through beds of thorns, cast him among slimy reptiles, into caldrons of boiling water, or beds of burning coal. Such are the sufferings of those who live ignoble lives, as fabled in the savage mind — fabled tortures, but tortures which fable only can convey. Fire, with its excruciating pain, can but feebly represent the anguish of a mind swayed by passions.

The spheres above the earth, where the spirits of the good ascend, or the Paradise of India, elicit the lavish encomiums of Hindoo poets. Those who are charitable and zealous in doing good ascend to the first sphere, above which there are various degrees of holiness, to the fifth, or sphere of Vishnu, where martyrs ascend. The sixth, or sphere of Brahma, is only attained by those who never speak a falsehood, and by widows burned on the funeral piles of their husbands.

Again, we remark the mingling of rich veins of truth among the fables, by which a heated imagination sought to set them forth.

Almost all the rhetorical figures employed by Hebrew chroniclers were used by the poetical Hindoo thousands of years previously. His Paradise is all the heart could desire. The sky is the softest cerulean, the waters of the clearest crystal, and umbrageous trees and odorous flowers perfume the air. The most enchanting melody is made by the Spirit of Music and the Singing Stars. The floor is of gems, and the pillars of the temples are precious stones. Godlike saints and beautiful women wander through the groves, beneath the silver light, devoting their time to contemplation and sacrifices to the gods. INDIA, the presiding god, is seated with his wife on an ivory throne, by the side of a beautiful lake, covered with lilies and lotus blossoms, her countenance beaming like a gleam of lightning, and her beauty filling Paradise with the odor of a thousand flowers.

The mythology of India was transplanted into Egypt; but so dim are the records of those early times, that the time and manner of its transference are wholly lost. The analogy of beliefs is very striking; so much so, that the question is settled in the minds of the learned that Egypt was colonized from India, or, at least, the dominant sacerdotal element of the nation was Indian. New gods were added to the calendar, new forms of worship instituted, but the central idea of immortality was not lost. It remained amid all its puerile trappings, and ever exercised omnipotent sway.

There were two principal sects among them; one believing in transmigration, the soul passing by successive stages through every being of earth, water, and air,—a circle it completes in three thousand years,—and then again reëntering a human body. They supposed that it would enter the same body that it left; and hence the extraordinary care they exercised in embalming the dead, and the enormous expense they lavished on mausoleums.

Fully impressed that the world was evil, and this life a probationary state, in which they resided as a punishment for crimes committed in a preceding, they looked forward with delight to the tomb, which they called their eternal home. Their sacred writings, as well as those of all other nations, abound with instances of communications between sainted men and superior intelligences, who ruled over particular objects and persons. They had oracles from which they received prophecies; and from the fame they acquired, and the confidence placed in them, they could not have been tricks of conniving priests. Those who aspired to receive communications from the invisible world fasted and prayed, that the spirit might obtain an ascendency over the body. This is the philosophical method by which clairvoyant perception is obtained, and in it we can easily perceive the source of the superhuman knowledge sometimes manifested by these oracles. It is also remarkable that the prophecies were delivered by females, in whom the clairvoyant faculties are generally larger than in man. Such was the confidence placed in these, that they were consulted on all important occasions, and most implicitly obeyed.

Beneath the same Oriental sky, remarkable as the cradle of mankind, but of another race, the colossal proportions of the Chinese sage towers gloomily through the mists of time. The religion of China is referable to Confucius, who lived five hundred and fifty years before Christ. He was one of the few genii who arise in the infancy of races, and with sudden stride leap across the abyss of ages, carrying whole nations with them. He is deified, as all great men are; and from Thibet to the Yellow Sea, from the Himalayas to India, the hills of the Celestial Empire are dotted with his temples. The books he compiled are the sacred books of the Chinese. In them occurs this remarkable passage: —

"How vast is the power of spirits! An ocean of

invisible intelligences surrounds us every where. If you look for them you cannot see them. If you listen you cannot hear them. Identified with the substance of all things, they cannot be separated from it. They cause men to purify and sanctify theii hearts, to clothe themselves in festive garments, and offer oblations to their ancestors. They are every where, — above us, on the right, and on the left. Their coming cannot be calculated. How important we should not neglect them!"

He taught that every individual was attended by a guardian spirit, who watched over and protected its charge; and the Chinese have images of these hung up in their houses, and worship them with oblations.

Contemporary with these was Persia and Chaldea. Their religious teachers, the Magi, explained the sacred books of Zoroaster, and possessed the power of prophecy. They taught that the human spirit once had wings, which it lost by its connection with the body, but which it would regain before reaching the celestial regions. Every individual had a guardian spirit to protect him from evil. The good and bad actions were believed to be written down, and every soul punished according to its deeds, ascending into spheres of happiness, or descending into a gulf of woe. They supposed that the inferior spirits could communicate with man, and that certain individuals, by consulting the sacred books, and by holiness of life, could approach near to the gods, (superior spirits,) and communicate with them. They understood by a holy life a rigid regime, and habits which would develop the spiritual perception, at the expense of the physical system, — the proper course to develop clairvoyance.

The wise men of the ancient nations understood the mesmeric art. From immemorial time disease has been cured by laying on of hands, and clairvoyant vision occurred.

In Moses' time, the Egyptian priests charmed the deadliest serpents, rendering them harmless and perfectly obedient, and perfected the art of magnetic influence to a great extent, ever concealing it, however, from the people's view by mysterious rites. Egypt was ancient in the youthful days of Greece, and Rome came after the decay of Greece; but the mantle of Egypt fell on them, and they treasured and improved her knowledge. In their history we learn more of the existence of spirits, and their communication with mankind.

" They believed that departed human spirits lingered around their former localities, and families, to protect them. They invoked them in time of domestic trouble, and offered sacrifices to appease them when they thought they had been wronged, or were angry. They erected costly tombs, and at stated seasons repaired thither to offer prayers and offerings to the spirits of departed ancestors, whom they called manes."

They at length erected splendid altars, and offered sacrifices to them as gods. If a man was a public benefactor, it was natural for the people to carry offerings to his tomb; and thus began hero worship. The spirits of departed heroes were supposed to become intermediate between mortals and the great gods, blessing the nations or individuals whom they protected, guiding their feet from evil, and filling their souls with great and noble deeds.

Hesiod, one of the most ancient of the Greek poets, records this belief: —

"Thrice ten thousand holy angels rove
This breathing world: the immortals sent from Jove,
Guardians of men, their glance alike surveys
The upright judgment and the righteous ways:
Hovering they glide to earth's remotest bound;
A cloud aerial veils their forms around."

The Hindoo idea of a subtile, invisible body, confined in the external, physical body, was transplanted into the Grecian mythology. It taught that man is composed of these elements — the soul, the invisible body, and the physical body. The invisible body was the tenement of the soul, and was carried with it when it went to the delights of Paradise, or to suffer the penalties of its sins in Tartarus. After its sentence expired, it was sent back to reënter another body, more or less honorable, according to its sentence. The Elysian abode of the blessed exhausted all the metaphors of the poet. The day was always serene, and a soft, ethereal light rendered the scene enchanting. Majestic groves and beautiful gardens variegated the landscape. The River Eridanus flowed through banks of flowers, and on its scented borders dwelt heroes and sages, artists and poets. There they engaged in the pleasures which formerly delighted them. There friends met in social festivals; the husband met his wife, and children greeted their parents, very much as the spiritual philosophy teaches.

On the other hand, they threw together all that was terrible and repulsive in the description of Tartarus, the abode of the damned. It was surrounded by a river of fire and a terrible wall. Here those who had lived sinful lives were scourged by the Furies; or had huge stones suspended over their heads, ever ready to fall; or hungry wolves or vultures gnawed at their vitals, which forever grew again; or stood in water, enduring the pangs of deadly thirst, yet unable to obtain a single drop to cool their parched tongues; or starved while delicate fruits were suspended just above their reach.

Some souls, too good for Tartarus, but too bad for Paradise, wandered in vast forests, exposed to scorching winds, until purified; others were plunged in deep water; and others were obliged to pass through intense fire to obtain the same result. If they were purified by this process they ascended to the gods; if not they were sent back to the earth to assume again the mortal form, and pursue another probationary period.

The Greeks divined the future by observing mysterious rites, and by direct inspiration. Inspirational prophecy was uttered by persons who were believed to be possessed by spirits; and while unconscious, motionless, and speechless, the spirit spoke out of their breasts. It was also made by persons who were seized with a sudden frenzy, or enthusiasm; and by those who fell into a trance, and when they awoke spoke of what they saw. Music was often employed to excite the prophetic frenzy, and it is well known that the succession of harmonious sounds on sensitive nerves is highly promotive of ecstasy, or clairvoyance.

Cicero says, "They whose minds, scorning the limitation of the body, fly and rush abroad when influenced and excited by some ardor, behold things which they predict. Such minds, which inhere not in their bodies, are influenced by various causes." The testimony of such a man has great force, and proves that spiritual intercourse has existed in all ages. But we rely not on his testimony alone.

It is said that Epimenides, of Crete, had power to send his soul out of his body and recall it at pleasure. In modern times he would be called a clairvoyant, or medium. During its absence he was as one dead, cold and inanimate. He frequently held intercourse with the gods, (superior spirits,) and was counselled by them. When a terrible plague devastated Athens, its citizens sent for him. He came, and erected altars to the Unknown God, and probably by exciting the religious enthusiasm of the people, by mesmeric and medicinal aid combined, arrested the farther progress of the disease.

Of Hermatimus, a famous prophet of Clazomenæ, it is recorded that his soul left his body, and wandered into every part of the world. While thus entranced, his wife, supposing him dead, had his body burned, according to the custom of the country. So much was his wonderful gift of divination prized by the people, that they erected a temple to him, and paid him divine honors.

Cassandra, princess of Troy, when a little girl, was playing with her brother in the vestibule of Apollo's Temple, and tarrying too late to be carried home, was put to sleep on a couch of laurel leaves. From that time she could continually hear the voice of the gods, or of spirits, who, by the ancients, were considered gods. She constantly foretold the destruction of Troy, and warned her countrymen of the stratagem of the wooden horse. She also foretold the manner of her own death, and of the Grecian conqueror who carried her away.

The wife of Paris, Œnone, is said to have had the gift of prophecy, and to have discovered the medicinal properties of plants. In other words, she was a clairvoyant, healing medium.

CHAPTER II.

PROOFS OF IMMORTALITY DRAWN FROM HISTORY, CONCLUDED.

The Roman Sibyls. — Oracles of Delphi. — Selection of Pythia. — Dodonian Oracles. — Brutic Oracles. — Pythagoras. — His Doctrines. — Socrates. — His Teachings. — Platonism. — Biblical Records. — Christ. — Early Church Fathers. — Witchcraft. — The Solution of the Problem by the present Age. — A new Argument drawn from the Nature of the human Spirit.

History is very ambiguous concerning the Roman Sibyls, a name bestowed on certain women supposed to be inspired by the gods. They fell into an ecstatic state, and were supposed to communicate directly with them. The most famous was the Cumæan sibyl, said to have written the Sibylline Books, which were consulted on all momentous occasions, and were considered as giving positive answers to all questions of state.

The oracle of Delphos was famous throughout the whole civilized world. Some shepherds were pasturing goats around the site of this temple, when they observed that when they put their heads in a certain place they ran and leaped wildly about. When the herdsmen did the same, they raved like madmen. The news of this miraculous grotto spread rapidly: a seat called a tripod was erected over the fissure, and a woman, chosen by the priests, was placed there during one month, in the spring of the year, to receive the inspiration, and answer those who came to consult the oracle. Lawgivers came to learn the most beneficial course to be pursued with their people; kings came to learn the fate of wars; individuals came to consult on the affairs of life. From the magnificent gifts of those who received benefit, a splendid temple was erected, and adorned with the most costly ornaments. This temple was situated on the south side of Mount Parnassus, and on the eastern side welled the Castalian fountain, in which the Pythia, or priestess, bathed before she approached the tripod. She crowned herself with laurel, and ate some of the leaves. As soon as she inhaled the vapor from the cavern she grew pale, her eyes sparkled, and she trembled in every limb. The priest attending, wrote down the words she uttered in her frenzy.

The effects of the vapor from the cavern might seem supernatural to the ancients, but in the light of modern science it is readily explained on natural principles. All narcotics, by debilitating the muscular system and unnaturally exciting the nerves, in a greater or less degree awaken the latent sensitiveness of the nervous system. In general this cannot be turned to a good account, disease following it so closely; but with Indian hemp, or hesh-sheiste, and exhilarating gas, it is otherwise, no permanent debility following their moderate use. Exhilarating gas produces the most startling effects; and undoubtedly a vapor very similar escaped from the cavern, and was breathed by the priestess of Delphos. The inhalation of this gas produces almost precisely the symptoms recorded of the Pythia, and were it breathed by sensitive persons, the symptoms would be identical.

Delphos was noted for the ambiguity of its answers, while Delos was famed for the directness and conciseness of its replies. Delphos was most famous, however, and the most ancient, being founded twelve hundred years before the Christian era. So infallible were its predictions deemed, that it became an adage — " As true as a response from the tripod."

The selection of the Pythia was intrusted to the priests, and with the practical mesmeric knowledge they possessed, they of course selected the most impressible persons they could find; and it will be readily seen that the value of the predictions depended on the degree of impressibility of the Pythia — an inference supported by the fact that though its truthfulness was always admitted, it was also confessed that this varied from time to time, sometimes being remarkable for distinctness and truthfulness, and at others equally so for its ambiguity.

Dodona was the most ancient of all the oracles of Greece. It dates back fifteen hundred and fifty-eight years before Christ. The oracles were delivered by a priestess whom Herodotus supposes to have been brought from Egypt.

The truthfulness of these oracles may be disputed, but they are as well authenticated as any portion of ancient history. Some of their responses silenced at once the charge of deception. So startlingly accurate were some of these, that a noted historian, unable to account for them in any other manner, refers them to the agency of the devil.

Crœsus, wishing to consult the oracles, first desired to test their truthfulness, and sent a messenger to seven of them, asking what was his employment on a certain day of the month. Designing to be employed in an occupation least liable to be conjectured, he cut in pieces a tortoise and a lamb, and boiled them together in a brass vessel. The Delphic Pythia sent him as answer,—

> "I count the sands, I measure out the sea;
> The silent and the dumb are heard by me.
> E'en now the odors to my senses rise —
> A tortoise boiling with a lamb supplies,
> Where brass above and brass below it lies."

Satisfied that the oracle was truthful, he presented his inquiries — first, whether he would be successful in his war with Cyrus the Persian, and as to the duration of his kingdom. Her reply was, that his kingdom would stand until a mule ascended the Persian throne; and when he crossed the river dividing his territories from the Persian, a great kingdom would be overthrown. He interpreted these answers as favorable to himself, prosecuted the war, and was soon overthrown and taken prisoner. Indignantly he sent a messenger to rebuke the oracle, but received the very soothing reply, that Cyrus, being half Mede and half Persian, was the mule referred to; and when they said a great kingdom would be overthrown, it was not by any means the Persian that they meant, but his own; and hence the prediction had been fulfilled to the letter.

The Brutic oracle told Cambyses he would die in Ecbatana. Supposing it to mean the great city of Media, he carefully avoided that place. Years afterwards he was suffering from an excruciating wound, and stopped to rest in an Assyrian village. Feeling that he should die there, he inquired the name of the place. They told him Ecbatana. The prophecy was fulfilled.

The Emperor Justinian had frequent intercourse with divine beings. They awoke him from slumber by touching his hand or hair. He knew them so well that when they came he could distinguish the peculiar intonation of voice of each.

Pythagoras was one of the wise men of Greece, born five hundred and eighty-six years before Christ. He taught that man was composed of an immortal mind, which was a portion of the Divinity, and had its seat in the brain; a sensitive, immaterial spirit, the seat of the passions, and a natural body, which the soul assumed as a temporary garment. At death the spiritual portion was conducted to the regions of the dead, to be happy or miserable until sent back to the earth to inhabit a new body. When purified by successive probations, it ascended to the regions of the stars, which he believed inhabited by spirits. He professed to hold direct communication with immortal beings, and to have visions. If tradition speaks truly, he possessed extraordinary magnetic power, and by it could make animals and men obey him.

Socrates was another sage. He makes frequent allusion to a demon or angel, which ever attended him. It accompanied him from his youth, and he says it never spoke otherwise than truthfully, and he always obeyed the warnings of its divine voice. He says, "When I was about to cross the river, the usual demonic sign was given me; and whenever this takes place, it always prohibits me from accomplishing what I am about to do. In the present instance I seem to hear a certain voice which would not suffer me to depart. I am therefore a prophet, though not a perfectly worthy one, but such a one as a man who knows his letters indifferently well — merely sufficient for what concerns himself."

It was a current doctrine in Greece that every man had a guardian spirit or genius; and the more friendship existed between the person and his genius, the happier and greater he would become. In other words, the more he cultivated his impressibility, the more knowledge would be given him from the celestial sphere.

The doctrines of Plato were similar to those of Pythagoras. He is represented as saying, "The soul of each of us is an immortal spirit, and goes to the gods to give an account of its actions."

Volumes might be filled with instances like those we have introduced, all substantiating the claims of ancient spiritualism.

The Bible records some of the most startling manifestations. In those primitive times angels were seen, and conversed with men. Saul's consultation with the witch at Endor is one of the most characteristic spiritual manifestations. She knew him not until she entered the superior or clairvoyant state. Then Samuel the seer appeared, and stretching aloft his airy arms, denounced him with awful, prophetic voice. She who was called a witch possessed high clairvoyant power, and the same is true of all the prophets of the past.

Looking far into the misty past, we are too apt to refer every thing we do not understand to trick or delusion; but when we consider the implicit faith placed in the ancient oracles, not only by the ignorant, but by the wisest philosophers, sages, and lawgivers, it is preposterous to suppose that they were entirely deceptions. False prophets are frequently mentioned, and their existence proves the true coin. The office of prophet conferred great honors and emoluments, and it would be strange if the temptation proved not too strong, and designing persons did not attempt the part of true oracles.

Clairvoyance has ever been possessed but by a very few, while the call for prophets has been constant and universal. But there has always been enough of the true to preserve the confidence of mankind.

The dawn of the present era beheld many startling spirit manifestations, which firmly support the spiritual philosophy.

The life and death of Christ was invested with spirit manifestations. Spirits appeared and conversed. His disciples meet, and suddenly a great light is thrown around them, a divine flame is on every tongue, and the poor, despised, illiterate fishermen of the shores of Galilee surprise the strangers from widely remote countries, by addressing each in his own tongue. They lay their hands on the sick, and they are healed; on the blind, they see; on the lame, they walk. Some are cast away on an island, and astonish the people by shaking off poisonous reptiles which fasten to their hands, and while they are expected to drop down dead, cure the sick by a touch.

The early fathers worked similar wonders, and were all impressed that the air was filled with invisible spirits, both good and bad. Clement, Apollonius, Ignatius, Polycarp, Justin, Tatian, Tertullian, Origen, Cyprian, all testify positively to the existence of spiritual intelligences. Origen believed that the spirits of the just went to the throne of God, and that they might by prayers and intercessions redeem those they loved on earth. He believed that man retained all his faculties and desires after passing the shadowy gulf of death.

Even down to the present, many believe in witchcraft, incantation, and foretelling the future; and so deeply rooted is this belief, that it is impossible to eradicate it. What mean the persecutions of its devotees? Have they been destroyed simply from a foolish effort to deceive? There is a truth somewhere beneath all this rubbish — a great truth, easily extracted.

In very recent times an extensive persecution arose from this cause. The Salem massacre blots the page of American history. It is evident that the individuals connected with that tragedy were ignorant of the cause for which they were deemed guilty.

So has it ever been in the world's history; a miraculous power is ever at work behind the moving canvas of human affairs, and here and there, only, crops out, like granite peaks from mist, revealing the deep force concealed. Whenever it has appeared it has been considered supernatural, and mistaken for a direct manifestation of God.

To the present age has been reserved the honor of solving this vexed problem. The analytical philosophy arranges the facts brought by history, and is delighted with their harmony. It sees one great law pervading the entire mysterious domain. To one force all facts are referable. The sublime philosophy of mind and spirit, which but to-day has been advanced, solved them all. That law is the impressibility of mind by which clairvoyance, in all its phases, and spiritual intercourse are maintained. History shows that both of these are very ancient. They were ill understood; and when a spirit spoke, its voice was considered as emanating from the gods.

True, it may be objected that we are not warranted in referring the historic facts to clairvoyance and spirits; but where is the room for mistake? We refer them to those sources as any effect is referred to its cause. It is an admitted fact that one mind can control another, and that some minds can pass into the clairvoyant state without assistance, in which state they can read the past and predict the future. If mind is thus susceptible to-day, it is probable that it was so a thousand or four thousand years ago. And when we find an event of the past, transpiring in precisely the same manner and under the same conditions as one at present, is it logical to refer both to a common cause?

The priestesses were nervous subjects; and when they entered the prophetic state they exhibited the livid, deathlike complexion, the contortions and rigidity of muscles, so well known to those who have investigated animal magnetism. It would be singular indeed if mind became susceptible to magnetism but a few years ago, after having remained unsusceptible for ages, and equally remarkable if the faculties were not made available. As steam was known to the ancients without their applying it to any useful end, so did they know the existence of mental impressibility without comprehending its vast importance.

For similar reasons do we refer many of the mysteries of the past to spiritual intelligences. If man is immortal, and retains his consciousness after death has consigned, with rude hand, his body to the dust, he must desire to come back and converse with his friends on earth. If the spirits of the good can look down from their celestial heights with calm indifference, and never desire to communicate the light which alone can dispel their earthly brothers' doubts, they must have lost the humane feelings which constitute the great and benevolent soul.

The Christian world unanimously believes in a future state. We will not stop to prove this step in our reasoning. Admit that there is a future state, what must necessarily be its characteristics? The immortal spirit, freed from the body, must be the same as it was in the body; with all its emotions and desires the same. Does not the father in a distant land desire to converse with his absent child? Does he not cross oceans and continents, breasting storms and dangers, to clasp him to his breast? How yearns the mother's susceptible heart for her absent son? Can the love of that father or mother be blotted out by death if he or she retain individuality beyond its shadow? Can friendship and the holy conjugal emotions become extinct? If not, then will the freed spirit, roaming among the bowers of the blessed, think of earth and loved ones toiling here, and, forsaking the pleasures of paradise, wing its swift way to earth, and hover around the loved.

This is the philosophy of the belief in guardian angels, which has existed in the world from immemorial time; and the host of genii, the ancients believed, overlooked the affairs of cities and nations. How plausible, that the man who shed his blood in defence of his country should, after death, retaining the same thoughts and desires, remain near to watch and protect it! Thus these myths of the past have a sound basis, and are not all vagaries of the imagination, as they seem. Beneath the most fantastic and grotesque forms of mythology ofttimes the grandest spiritual truths are concealed.

But these guardians would be engaged very unprofitably if they could not communicate with or influence in the least those they guarded. Nature never suffers such an imperfect arrangement; and so certain as there are guardian spirits, so certain they communicate.

The following propositions rest on the admission of man's existence after death:—

1. If he exists, he must retain all his ideas, thoughts, faculties, desires, and emotions unimpaired.

2. If he retains these, he will desire to commune with those he loves on earth.

3. If he becomes a guardian spirit over those he loves, he must have some avenue through which to communicate with them. This reasoning is not only true for the present, but for all past time. It is inherent in the constitution of man; and though we may suppose impressions were made with more difficulty on the crude and undeveloped mind than at present, yet that they were produced on the most susceptible, facts previously stated conclusively show.

We are born into a world of which we at first know nothing. Above and around us spread the clouds and the sunshine. Above us nightly watch the silent stars, and around us is the activity of animate nature. Through all these, the soul develops step by step, until at last it feels the mighty power within, proclaiming its own divinity. If the soul is immortal, it should know it. The great consciousness of its existence should dawn like a divine radiance, and fill it with inexpressible hopes and aspirations. Hence this universal consciousness we have shown to exist is a strong philosophical argument. It is not educational, it is not imbibed, for there must have been a time when it was not known. From whence came this knowledge? Was it whispered by the spirits of the departed into the ear of the savage as he lay pillowed beneath the waving trees? or was it the dim and undefined aspiration of his own spirit for the Great Unknown beyond the Lethean flood of death? We would say, the aspiration of his spirit, questioning itself — answered by itself.

Thus do we see that, amid all the vicissitudes of time, one great and fundamental belief has pervaded the heart of humanity. On the immortality of the spirit the theologies of the world have rested. Above the sphere of mortal affairs, in the clear ether, the spirits of the dead exist, and frequently communicate with their earthly brothers. They were the mediators between the unapproachable holiness of the Deity and man.

Amid the labyrinths of mythologies and theologies of the past, the great law previously adverted to — impressibility of the mind, or mesmerism — pervades, working supernaturally, and astonishing the nations by its miracles. The vast and apparently disorderly fabric is reduced to system and order, and from its confusion, reason, pursuing the guidance of inductive philosophy, builds a temple whose foundations are on the earth, but its spire pierces the veiled heavens of the spirit-spheres.

4*

CHAPTER III.

EVIDENCES OF MAN'S IMMORTALITY DERIVED FROM MODERN SPIRITUALISM.

The Method by which we propose to make our Revelations positive. — Proofs: — Moving of Tables and other ponderable Objects. — Intelligence manifested. — Laplace's Problem of Probabilities. — The Chain of Arguments, Objections, and Theories considered: — 1. Are Spirit Manifestations the Work of Satan? — 2. Of Evil Spirits? — 3. Are they produced by detached vitalized Electricity? — 4. By Od Force? — 5. By Deception? — By Hallucination? — Identification of a Spirit: — Identifies the Individuality of all others. — Varied Forms of Communication. — Object of. — Our Evidence becomes positive.

We subject ourselves to the humiliating task of proving our own identity before applying ourselves to the main purpose of this volume. In doing this, we shall treat our subject as we should were we of earth, and attempting its substantiation by positive testimony. If we accomplish this purpose, we make our words as positive and conclusive as the preceding volume, where direct facts were adduced. Before the witness is heard, it must be ascertained whether he be trustworthy, and what he purports. If we meet this test, our description of the spirit world becomes as positive as the narrative of the traveller; for we are guided by our senses, and write as they teach us. For this reason, we have not only opened this volume with a dissertation on Ancient Spiritualism, but follow with a summary of what we consider positive proofs of the truth of present manifestations.

If one spirit can be identified, the proposition is proved; for the method that will identify one spirit will identify all others. A spirit visits a circle, and moves the table, the chairs, or elevates the medium or members of the circle above the floor, doing so without any visible contact or agency whatever. This is admitted, and it is unnecessary to quote facts in support. A force is exhibited, a mysterious force, which received science has not, cannot account for.

So far individualized intelligence is not manifested. The force may be produced by magnetism, Od force, or any other of the unintelligent agencies of nature. Of all this class of phenomena, we will not pause to dispute, as they are unessential in establishing our position; for when the main point is admitted, these will readily fall into place.

It is true, no intelligence is manifested; but mark, a question is asked — the table moves in response. The answer is correct. One question may be answered correctly by chance. A happy coincidence may give the second answer correctly, and even the third, we will admit; but how stands the case when a hundred successive questions are answered correctly? Now, the question of intelligence resolves itself into Laplace's problem of probabilities. Against a single question being answered correctly, if the force which moves the table is not an intelligent force, or backed by intelligence, there is an infinite improbability if the answer is not yes or no. What, then, can be said when a hundred consecutive questions are answered correctly, without a single failure? The chances of error are reduced to nothing, and in a mathematical sense the proposition is established.

From whence is this intelligence derived? from the mind of the circle, or medium? It may be when the answers are known to them, which for the argument we admit; but there are instances where the answers are unknown to any one present — answers which only one being knows, and that being is in the spirit-world. How strong the inference! Such is the conclusion derived from physical manifestations. They present themselves in the following chain: —

Physical matter is moved by an invisible force.

This force is not derived from the circle, the medium, or any human agency, as it is superior to human intelligence.

There is infinite improbability against its being any thing else than what it purports.

The intelligence manifested is not mundane.

It identifies itself with the individual from whom it purports to emanate, by answering questions which that individual only can answer.

We have not completed the extensive list of manifestations. We write, we impress, we speak, we produce trances and visions; we appear under favorable conditions, show a phosphorescent hand, or directly touch the person.

In answering questions through any of these channels correctly, the doctrine of probabilities can be usefully applied.

In all these manifestations objections arise, which we will briefly consider.

1. It is the work of the Devil.
2. It is Evil Spirits.
3. It is Electricity, " detached and vitalized."
4. It is Od force.
5. It is Deception, — a cheat.
6. It is Hallucination.

Of all these theories and objections, not one is worthy of a moment's notice. One after another has been exploded, with countless others of lesser fame, until, weary of repeated failures, their sapient propounders have ceased to promulgate them. However, we will give each a brief review.

1. The tree is known by its fruit. Our communications in general have a high moral tone, which internal evidence alone establishes their claims, and negates the supposition of their diabolic origin.

2. There are spirits — not *evil*, but degraded, miserable spirits — who communicate with earth; but the same law which permits them, also allows the good to converse. Ah, ye Christians, who promulgate this theory, what can be your idea of the just and good Deity you worship, who allows the myriads of the damned to deluge the world with their contaminating presence while he debars the counteracting influence of the good? Forbid it, Heaven! We trample the idea beneath our feet with scorn. Law never works in opposite ways. The road which allows one passenger to go, allows all. Good and bad alike visit you, and communicate their best thoughts. You have reason. Judge for yourselves.

3. Until we are told how electricity can be detached and vitalized this assumption is worthless. Electricity has no more intelligence than water. Whence, then, is the manifested intelligence derived? and more, in and around the article moved, not the slightest indication of this agent can be detected.

4. The claims of Od force are still less worthy of consideration. It is forced into a position of which its discoverer never dreamed, and given power which

in all his investigations he never detected. The flames which play around the poles of magnets and crystals, only detected, so thin are they, by the most sensitive nerves, — what have they to do with moving ponderable bodies, or suspending them in the air?

5. Mediums may deceive; circles may be humbugged. We admit, for argument, that in nine cases in ten they are; but the tenth we will not admit. When the medium is, by invisible hands lifted to the ceiling, and suspended there; or when the table is elevated in the same manner; or when questions are answered which no one on earth, not even their propounders, can answer, — we hold that the supposition of deception is puerile.

6. It is impossible to hold the position that all these phenomena can be accounted for by supposing the spectators are hallucinated. If so, then the real world melts and fades into a dream, and there is no reality in any thing. If the senses are not to be trusted here, why any where? If the eye sees what is not, the ear hears what is not, the touch feels what is not, then, observation and boasted scientific accuracy, good by. The world becomes a magnificent phantasmagoria in which we dream, but are not. The German mysticism becomes realized — we think nature exists, that we exist; but it is all a thought, a juggler's delusion, for there is nothing in space but a void.

Our manifestations have been varied as theories arose, until now we have thrown our opposers into a very unfortunate dilemma, for whatever explains our facts sweeps away all the supernaturalism of the past.

So varied, so numerous, so common have our communications become, that the introduction of facts would only encumber pages we propose to fill with other matter.* We only attempt to roughly sketch the main argument, and plainly state conclusions derived from facts elsewhere stated. The identification of a spirit decides the controversy. How would you identify a friend concealed by a wall, if you could not recognize his voice? You would ask for proof that he was really the one he purports to be, and he would tell you some incident, some sentence peculiar to him, and him only. We are concealed by the wall of invisibility. You ask, Are you whom you purport? We answer with some familiar sentence. Is not the identification perfect? There is an individuality acknowledged; our responsibility and trust allowed, and our revelations become positive knowledge. Such are the views with which we commence this volume; and though we shall ever argue the questions under consideration, many will not admit of more than a simple narration, as the description of our homes, &c., which must be received on our word.

* Several volumes of facts have been compiled, which are accessible to the inquirer, and to which he is referred. — Night-Side of Nature, (*very valuable*); Hare's work on Spiritualism; Owen's Footfalls; the Telegraph Papers; The Shekinah, &c.

CHAPTER IV.

THE OBJECTS OF MODERN SPIRITUALISM.

Position of Christianity. — Jewish Religion. — Of Christ's Reformation — Revelation. — Progressive. — Not infallible. — Mutual Relations of Revelation and Science. — State of the World. — Impossibility of believing what is contradictory to Reason. — Tolerance. — The Combat between the Conservative and Reformer. — Primary Object of Spiritualism. — Mistaken Ideas. — Spiritual Beings the true Philosopher's Stone. — Warning Man of Danger, discovering Treasures, detecting Crime. — The Truth declared. — The true Object.

Christianity has taken one step, for it is progressive. The new dispensation supplanted the old. The Jewish theology answered the wants of an early and savage race. Hard as iron, inflexible, and bloody, it was *the* religion for the Jew, the most cruel and bloodthirsty tribe of ancient days; so brutal and avaricious, that he was as universally hated as despised.

His Jehovah was an enlarged view of himself. His religion was like his God. It assumed the arrogance of bigotry, and declared a handful of savages God's chosen people, authorized to slay and mangle their enemies whenever interest dictated; to sack cities, and butcher thousands, God thundering his assent, urging them on to carnage, and participating until his garments were red with slaughter. Such a religion, bad as it appears, was the best religion they could comprehend; and no better could then be given them.

After some thousand years, Christ came, and gathering all the best ideas of the ages, advanced to the outposts of thought, and began a new dispensation.

Pure as the crystal stream when it flowed from him, we know how sadly it has been corrupted, until it no longer slakes the thirst of the present. Does it teach enough of God, of immortality, of the true life? Let the aspect of the world, its wranglings and discord answer. A perfect revelation does not need an interpretation, nor an explanation; for it is a clear enunciation of the truths of nature. It is characteristic of imperfection to be misunderstood, to require notes and explanations. We do not expect perfection in the world of mind, but we expect a more perfect state than existed two thousand years ago. As the race progresses, new and higher revelations are received. Revelation is progressive, and if it sets its landmarks, is certain to be-become an encumbrance. As mind advances, new truths flash out along its path, sending their illuminating rays into the past, and penetrating the darkness of the future. Science — the classified knowledge of the race — moves slowly onwards, and the so-styled revelation has given way before it. One strong position after another has been surrendered; here a grim castle, there an impregnable redoubt, walled by superstition, has been evacuated, for science hears no capitulation or compromises; it heeds not the voice of any book, be it never so old or sacred. With inquisitive eye it pries into the mysterious, the hidden and obscure, and boldly enters the most sacred domain. With rude hand it takes down the holy volumes of the nations, and reasons on their words with cold impiety.

All through the thick folios its explanations disagree from the word. How shall the truth be known?

Which explanation has been received? Which taught in the schools? Invariably the facts of science.

Science is ever stern and inflexible. It never retreats from its positions, while it has forced the so-styled infallible revelations to bend before its invincible evidence.

Glance for a moment at the efforts which have been made to reconcile the first chapter of Genesis with the geological account of the creation. Volume after volume has been written, but facts remain unanswerable. There can be no reconciliation. (See vol. i.) Something cannot be created from nothing; the world, instead of six days, must have been countless millions of ages in forming. One says that the rainbow is a symbol of God's covenant with man. Who accepts this explanation, or believes that no bow was painted on the storm billows before the flood? Nature has spoken, and her revelation is received. The prism forms an artificial rainbow, and thereby explains how light striking the falling drops of rain is refracted. Whenever the shower has fallen, the bow has girted its brow — ay, millions of years before the flood.

Who believes in a universal deluge, or that the earth is flat, and the centre of the universe, around which sun and stars revolve?

The revelations of nature have been received. The sacred books of the nations have answered their end, and no longer satisfy. When urged on the present they are failures.

Look abroad over the world. Do you see harmony, concord, peace? Rather the worst discord — of ideas and actions. Reform after reform is called for by those suffering under the grievous weight of existing institutions, while the old is left behind by reforms arising from its ruins. The new and the old have their advocates, who cling with fanatical devotion to their own systems, and listen to the claims of no other. The errors of the world cover it with a thick and impenetrable forest, which arises before the reformer like that which met the pilgrim's gaze along the Atlantic's frozen shore. They met with difficulties, but with indomitable energy grappled with the ruggedness of nature. They cleared away the forest, blasted the rocks, and drove the plough over the hills where the red Indian still chased the deer with his swift arrow.

Thus melts away the growth of error before the reformer. Trees of centuries' growth are to be felled; the soil ameliorated by plough and harrow; the rank weeds, ready to spring up, mown down; and when every thing is prepared, the seeds of truth sown with a strong arm, and they will bloom in immortal verdure.

The reformer meets a more savage foe than the pilgrim met. That foe is the time-serving advocate of the past, the conservative, who clings to the record of bygone ages, and shuts his eyes on the present. Fired with the unforgiving zeal of bigotry, they are less merciful than the red man gloating over his mangled victim. They build the gibbet, the rack, and invent horrid instruments of torture. They fire the fagots, and scoff, and sneer, and spit on truth nailed to the cross, and have ready the crown of thorns. See how it has ever been with the two classes — the one that moved onward, and the one that stood still. Has there ever been any tolerance manifested by the latter? What little tolerance there has been was with the supporters of newisms in their infancy, when they lacked power to force their doctrines on those who thought otherwise. As soon as they obtained power, however, they set down landmarks; embodied their belief in a creed; warred desperately against all farther progress, and sought to dam the waters of the river of life.

The waters collect. The brook and the river add their might, and the torrent and cascade rejoice to augment the flood. The shower and the storm add their might, and the stream which leaps from the glaciers' home away up among the mountain peaks, hurries on to join the accumulating force. The flood moans and hisses, in pent-up rage, while the anxious conservatives labor at their dam, stopping up seams and fissures which are constantly opening. Small crevices open at first, but gradually enlarge until the great stream roars through. Conservatism gives way. It cannot control the gigantic force which has slowly accumulated, and with tremendous crash, dam, conservative, and all are swept onward by the wild flood. Then there is confusion and disorder. There is no stability any where.

At length the force is expended, the whirlpools less violent, the roar less deafening; and again the conservative chieftains marshal their army, and away down the stream begin to erect another obstruction. Say they, "Humanity can never reach as far as this: There is no danger of a spiritual growth beyond this point. Here we can safely dam the progressive stream, else get too far, and prove dangerous." They begin the work by dragging out of the whirlpools the rotten timbers of the old structure, a plank, and a bolt, and a bar, and binding these together with new timbers, they exclaim, "Now we have bound this rebellious stream. Here is the *Ultima Thule*. Rest here forever.

All knowledge beyond this point is useless and dangerous, for here is the limit assigned by God for the human mind."

A little stream comes down meandering among trees and flowers, and where the wild rose and anemone perfume the air, and the loving birds hold concerts. It runs up against the barrier, then along its foot, endeavoring to go onward by some chink or crevice which thoughtlessly may have been overlooked. All is stern and forbidding. Every where a hand, made doubly careful by experience, has applied itself, and not the least flaw is perceptible. The waters gather, unite in little pools, and spread along the foot of the embankment. Then the floods run along the course of the meandering stream, until it becomes a vast river, which bears the obstruction away on its bosom like a floating straw.

Thus has it been with the world's progress. There never has been a sufficient accumulation of force to sweep away obstacles. The momentum becoming rapidly exhausted, the opposing force would gain time to recruit. But a reform is now ushered in which has the combined strength of the spheres to urge it on, and will sweep away every vestige of opposition.

To introduce a true and dignified rationalism is the prime object of spirits. To cut humanity loose from the fetters of superstition, and free reason from the chains of creeds, is a work sufficient to enlist the services of archangels.

Not so, however, has our mission been understood. A few visionary minds, at the dawn of spiritual manifestations, viewed the subject as an easy method of gaining a special explanation of the phenomena of nature; and they judged of its importance in proportion to its accomplishment of their own selfish and personal ends. The spirits have come, and sure we can have an explanation of what the philosophers have failed to explain. They expected, with easy method, to set themselves up as the philosophers of the age. Not a secret in heaven or earth but should be published. And this was the mission of spirits: They had come to stand by a table and rap out answers to selfish questions, and pander to the egotisms of querists. The inhabitants of heaven had come to earth for the special purpose of becoming lackeys and slaves; to stand for hours and submit to a cross-questioning as severe as prisoners at the bar. It was thought they would supersede the telegraph, and convey intelligence across continents and oceans much cheaper. The moving of a table, suitably applied, would be a cheaper locomotive power than steam. I need not tell you that this class of expectants were disappointed; their hopes utterly blasted, and with the genuine character of disappointed expectants they turned round and began a reviling as low and contemptible as their former vanity. "Why," ask they, "do not the spirits tell us the fate of the lost Pacific? Why do they not forewarn their friends of accidents? Why not tell us of Sir John Franklin, and unveil the horrid tale which is attached to his adventures in the Northern Sea? Why do they not guide our feet from dangers, and keep us in the path of success?" Ah, friends, spirits find more profitable employment than any of these. Are such the pursuits you hope to follow?

"Why not warn us of danger?" If we reason thus, where shall we find ourselves. We must have a spirit engineer on every train of cars, a spirit con ductor, and a spirit at every brake. We must have every steamship officered by them, and one at least to each passenger, as a watch to guard their footsteps. Then what use is there of any human engineer, conductor, or brakeman, as long as it can be done so cheaply? The latter would certainly be an important consideration! These could be dispensed with; and the passengers, all being cared for, can stumble as carelessly as they please, for they are in no danger, their footsteps being watched and guarded. Now, what use is there of man at all? Of what use would be his existence? He would become incapable of self-action, his identity lost, the mere instrument in the hands of a superior intelligence. Of himself he would be incapable of executing the slightest plan; the spiritual would be all predominant, and he but a puppet in their hands. Is not this revolting to manhood? Is it not better to have a collision now and then, a steamer lost, than to thus surrender personality? We have not come to perform tricks to delight men, or to use them as puppets with which to work. This is not the mission on which we come, nor the object of our philosophy.

We do not see why those who bring forward such objections as the above do not add the immense service which might be performed by having invisible marksmen to train their guns and turn aside the bullets of the enemy in battle, and enlist the spirit of Napoleon to order the combat. Then how the enemy would melt away before the heavenly hosts, who unharmed could fly over the battle field strewn with the dead and dying, and train the death-dealing cannon at the thickest ranks of the foe!

There was another class that thought they would make their fortunes out of the new philosophy. They would, by its means, discover some secret treasure, — a gold mine, — and become rich. The spirits knew, of course, all about their selfish matters, and would be willing to hunt up buried pots and kettles filled with eagles and dollars, and ransack mountains for placers. What excellent and knowing slaves they would make! What a glorious thing this spiritualism was! It made men rich! Without a stroke of labor they could procure gold, the instrument with which they could crush and fetter their unfortunate brothers. Yes, it would make them rich, and was *really* a glorious affair. They were excited about it. Who would not be excited when dollars could be made? They were sensitive, quite *impressible* — in the pocket, their most impressible part. The bubble burst. Lo! vain expectation! disappointed ambition! They found at last that the spirits did not come into the world to be lackeys for the amassing of wealth. They come not for the aggrandizement of the few, who in the beginning showed how unworthy they were by such selfish and egotistical questions. "If you would have wealth," they exclaim, in the sweet harmony of the spheres, "if you would have wealth, labor honestly at some useful toil. Do not ask us to help you, or to give you an instrument to enslave your neighbors. Then to the field, the workshop; get you thither, and become useful, and not a drone in this laborious hive."

What a question to ask an immortal being — How make dollars fastest? There are men now who can make them fast enough, quite to the satisfaction of every body. There are men now who can pinch it out of suffering and agony — pinch it until the eagle on it screams.

There was still another class, who saw in the advent of spiritualism the terror of all evil-doers. There would be no necessity for police or constables any longer, and old offenders would now be arraigned for their misconduct. What a glorious thing spiritualism was! No more trouble of protecting wealth; no need of safes, or iron chests, bank locks or Hobbs locks. Money would be safe lying about carelessly, night or day. If any one stole, the spirits would detect them. Then the expense would be greatly lessened — a great consideration; and men could get rich faster, which was a still greater consideration. Thieves began to tremble, they feared their doom was approaching. Then it was ascertained that God did not send his ministering angels into the world to be made watchmen or police, to guard a few petty dollars that an old wretch of a miser could leave unprotected, because he was too penurious to pay for the expense of a safe. "O, never!" speak the angels; "what care we if thieves get thy gold? Perhaps thou hast obtained it as wrongfully. We care nothing about thy wranglings. What are money and chattels to us? If they are taken from thee, art thou less a man? It will teach thee to be careful, and to exercise more discrimination. The thief must live — put him in prison yourself; we will take liberty away from no man. If thy dwelling is fired, see that thou provoke not thy brother in future, until he will do so foul a deed; it is no concern of ours. You are of the world, but we are spirits in heaven, who stoop not to share your wranglings."

This lesson has been learned by sad experience. We have none too much individuality. In the whole world, few there are who are individual in thought. This one has his identity swallowed up in some great name; that one in another. But if the little that does exist is to be destroyed, what a miserable set mankind will become! The assumption of such control, the becoming what we are requested to become, would be a direct and flagrant violation of the laws of man's being, and would be inevitably of the most ruinous consequences. There is an equilibrium established at present which is touched with danger, and to destroy which would unsettle the constitution of society. Yet this is asked for by such questions. They are all for self-aggrandizement, and meet a just rebuke. There is an instance, now and then, of lost money being discovered, and stolen goods restored; yet such are exceptions, and for the best of reasons. Whenever and wherever selfish questions have been asked, they have met a just rebuke. Slowly and practically this has been inculcated: that we came not to teach how men might obtain wealth. Our aims are immeasurably higher than the broadest coin. We have nothing in common with worldly wealth.

Property is good when rightly used; but cannot men get it without Heaven to help them?

Precisely the same expectations were aroused by magnetism and clairvoyance — the John the Baptists of spiritualism. How signal was the failure! These were unfortunate, however, in being easily accessible to quacks and charlatans, and passed a fiery ordeal, but have come out unscathed. If the hopes of the magnetizer nad been realized, there would have remained nothing for the spirits to discover. To them clairvoyance was the key by which the secrets of heaven and earth could be unlocked and demonstrated. The royal road to learning had, at least, been discovered, and they were a long way in the right direction. What was accomplished? A few quacks made pennies by it.

To ask such questions is insulting the good beings who bring to you the glad tidings of immortality. It reminds us of the messengers the lord sent to ascertain the condition of his estate, whom the laborers imprisoned and maltreated. So the good God has sent his angels into the world to take their brother man by the hand and point him heavenward; to whisper consolation to distress; to relieve the wretched; to agitate the great fountains of light, and stir up the latent energies of the soul, founding a new and harmonial philosophy, with man for the object and heaven for the aim. He never paused to consider the influence the dollar would exert on its spread. He trusted to the might of invincible truth. He thought only of the darkness which existed here; the gasping, dying condition of the soul for spiritual truth; the agonizing hope for more positive knowledge. He thought only of the religious controversies; the wrangling and disorder caused by ignorance; the superstitions which bowed angelic man down in the dust, and chained him there with creed and ceremony, stupefying his faculties with the opium and hemlock of bigotry. He thought of the dreariness of science without life; the errors of the learned; and he said to the good angels, Go down to the earth, and teach thy erring brother that ignorance is the mother of all vice; that there is but one religion, and that is philosophy. "O, stop this quarrelling! All are brothers. Have tolerance for each other's difference in belief. Are ye not all fallible?"

CHAPTER V.

CONSIDERATION OF SPIRITUAL PHENOMENA, AND THEIR. DISTINCTION FROM SUCH AS ARE NOT SPIRITUAL, BUT DEPENDENT ON SIMILAR LAWS.

Introductory. — Division of the Subject. — 1. Mesmerism; 2. Somnambulism; 3. Hallucinations; 4. Apparitions; 5. Dreams; 6. Influence of the Medium; 7. Influence of Conditions; 8. Position and Intelligence of the Communicating Spirit.

CREDULITY, which believes without positive evidence, is as reprehensible as unyielding scepticism. Both are puerile and ignoble. Human nature, however, inclines to one or the other extreme. It believes all or rejects all, and the fable of " strainers at gnats, and swallowers of camels " is ever enacting.

Many noble-hearted men have received spiritualism with unbounded credulity, and by the extent of the influence they assign to it brought the contempt of sceptics on themselves and the cause they advocate. By this course they weaken their defences, and the enemy finds multitudes of vulnerable points of attack. By making spiritual influence account for phenomena, closely related, but distinct, the opposition gains great advantage by proving the contrary, and then extending the inference to the whole domain by a trick of logic.

Such reasons have influenced us to devote a chapter to the consideration of spiritual phenomena, and such as resemble them.

Man is a spirit as much while in the body as out

of it, and consequently, as far as his corporeal state will permit, governed by the same spiritual laws. From this cause confusion arises, as there is a perfect blending of phenomena at the borders of the two states; so great is this confusion that we can safely estimate that one half of what are called spiritual manifestations are of mundane origin. Not that direct humbug is used, but mediums and circles are deceived.

In considering this subject, we shall, for brevity, divide it into, —

1. Mesmerism; 2. Somnambulism; 3. Hallucination, and Insanity; 4. Apparitions; 5. Dreams; 6. Influence of Mediums; 7. Influence of Conditions on Communications; 8. Position and Intelligence of the Communicating Spirits.

1. *Mesmerism* — Is the key to the spiritual philosophy, by which only it can be understood. *One spirit in the body can influence another spirit in the body.* Such is a general statement of the law of psychological influence. We shall, in its proper connection, show that we influence mediums precisely as the magnetizer does his subject, and that the body is nothing in the intercourse of spirits.

2. *Somnambulism.* — This is a state of mind very nearly allied to that produced by mesmeric passes, but is not induced by them. The mind enters it from organic derangement, and dream-land becomes actual. Volumes might be filled with facts, showing how, when apparently wrapped in profoundest slumber, the somnambulist has performed most surprising feats, such as climbing from a chamber window to the roof without aid, which would be impossible for any one

to perform when awake; getting into and out of positions which appear incredible. It is said that a lad in the Highlands scaled a perpendicular precipice to an eagle's nest, which had never been scaled before, and had always been deemed inaccessible.

In this state, which verges on clairvoyance, and sometimes is identical with it, the spirit is freed from the body sufficiently to possess senses of its own, and have no use for those of the body. In it, beautiful pieces of music have been composed, sermons written, and surprising mental operations performed. In pronouncing on such exhibitions care should be used not to confound the operations of the mind with spiritual influence, as has been rashly done, for it must ever be remembered man himself is a spirit, and capable of manifesting spiritual phenomena.

Still more caution should be used in the next class of facts. The mind remembers all the occurrences of its life. They may be dimmed on memory's tablet, but never effaced, and the proper conditions will awake them fresh as the occurrences of the hour. Sydney Smith experienced this when drowning. He says that all the events of his life, even to the most minute, rushed, in a minute, before him. This is true of all spirits, and it is this that makes the good action and glorious thought bestow reward, while crime and evil inflict punishment by ever presenting their horrid forms. Understanding this, we can readily account for those remarkable mental manifestations, when, during sickness or some derangement, the most trifling occurrences absorb the whole mind. An ignorant servant girl, during sickness, converses in the learned languages.

learned languages quite to the astonishment of her attendants. Now, this was not spiritual communication in *tongues*, but a natural result of mental derangement. Years previous she resided with a scholar, who often repeated these passages, and thus wrote them ineffaceably on her mind.

Another anecdote is told of an individual equally astonishing her employers by imitating the sounds of a violin. This was not a freak of a musically-disposed spirit, but the girl had previously dwelt with a musician, and in her sleep imitated the sounds which then had vibrated on her mind.

We would not be understood as denying that such spirit manifestations exist, but we deny that all such are supermundane. Mediums may be influenced to produce the most thrilling musical sounds, or converse in languages by spirits; but such phenomena should be rigidly examined, and never confounded with the operations of the medium's mind.

3. *Hallucination, Insanity, &c.* The transition to these is easy, and the danger of confusion and error still greater. The mind is so constituted that it is unbalanced by an undue determination of blood to any given region of the brain, produced by over exercise of that region. The evil goes on increasing, nature becomes unreal and fantastic, and it is impossible to predict where the unfortunate subject will be ed. In its mildest form it is hallucination, but rapdly strides to insanity and madness. Some have supposed that all insanity results from the influence of depraved spirits; but such is not its primary cause. Exceedingly impressible persons may be driven to madness by depraved spirits, or they may, in pre-

cisely the same manner, become so by falling into inharmonious relations, as by being surrounded by antagonistic persons and conditions. But we are not to consider spirit influence the primary cause of insanity.

Derangement of the mind, whether natural or produced, may furnish the requisite conditions for the control of inharmonious and degraded spirits, and these, by seizing the opportunity thus supplied, may greatly aggravate the disease.

We shall first consider the purely mental instances of hallucination, and rapidly approach insanity. Some friends are dining, when one receives a letter, bites off the seal, which he accidentally swallows. His friend exclaims, with affected horror, "It will seal you up." This produces such an impression that the unfortunate man refuses to eat, declares he is doomed, and presently dies. Did an evil spirit seize this favorable moment to destroy his enemy, or shall we refer it to a purely mental impression? The latter, unqualifiedly, in all such instances. In an insane asylum a person has been confined over twenty years. He labors under the idea that he is God. Volumes of such instances might be written. Hallucinations assume the most varied complexion, sometimes harmless or amusing, at others, when the animal faculties are excited, of a dangerous and alarming character.

Now let us pass to what are called hallucinations, but which are really spiritual influences. In this catalogue we shall place the oracles of all ages, the fantasies of their priests and priestesses, and the cases of real (not forged) witchcraft. History is filled with instances of such influence, which is classed under the general head of aberration of mind by philosophers. One must suffice for illustration, and the reader can examine the history of the world for countless others.

Almost at random we select that of Joan of Arc as an instance of the inhabitants of the spheres having mingled in the affairs of men, and by their influence decided the fate of nations and kingdoms.

The most wonderful tale of history is that of Joan of Arc, so wonderful that the world repeats the question of De Quincey, "What is to be thought of her?" The shepherd girl of Domremy for five centuries has attracted the attention of mankind. I give the details of her inspiration, because she illustrates the spiritual relations and nature of mind in a most beautiful manner. Her temperament was extremely sensitive and finely organized. She was modest and retiring. In her earliest childhood she held converse with spiritual beings, which she supposed fairies and elves, beneath an old tree by the banks of a little rivulet, which was consecrated as a place they loved, by the popular traditions. Her senses were so fine that she could see them, and hear exquisite music. In her thirteenth year she saw apparitions, and angel forms as bright as noonday. She was standing alone in her father's garden, when suddenly she saw a most brilliant and beautiful light shining into her face, and while almost overcome by the wonderful sight, she heard a strange but sweet voice, bidding her "be a good girl, and God would bless her." Her heart was pure and unsullied.

While alone with her flocks, another vision came. In the sky above her, wonderful and majestic forms floated, and she was addressed in mysterious language. Then it was told her that she should deliver France.

It is a singular propensity of the mind to endow the celestial being which most occupies the thoughts with the attributes of a special guardian. The Hebrews supposed that Jehovah had especial control of their nation, and the spiritual beings who guided their prophets they personified in him. In Greece the national gods spoke through the oracles in the same manner. So Joan supposed that the spiritual being she saw, and who guided her, was the angel Michael. When on trial, several years after, she said, "I saw him with these eyes as plainly as I see you now."

She says, "When the saints were disappearing, I wanted to weep, and beseech I might be borne away with them; and after they had disappeared, I used to kiss the earth on which they rested."

She was deeply impressed with her revelations, and as an omnipotent power seemed to endow her with the mighty mission of delivering France, she would not be turned from it by any circumstance. She devoted herself, soul and body, to her country. She rejected the matrimonial offer of a young countryman, because she felt the necessity of her remaining free. Against the wishes of her parents and friends, she set out for the court of the French king, penniless, and without any reference whatever, inspired by the irresistible destiny of her mission. Rebuff succeeded rebuff; but at last she impressed two gentlemen with her enthusiasm, and they conducted her to the throne. Still the king would not grant her audience. His courtiers questioned her. They were influenced by her enthusiasm, and recommended her to the king. He, doubting her truthfulness, seated a follower on the throne, while he, in plain dress, mingled with the crowd. He thus proposed to test the spirit she said controlled her, for if she was simply an enthusiast she would address herself to the throne, but if guided by a higher power she would single him from the crowd.

The modest, retiring shepherd girl was conducted into the august presence. She seemed to forget herself, and only remember her mission. She stood erect, and gazed around her; turning from the throne, she approached the king. "In the name of God," she said, "you are the king. I am Joan, sent by God to aid you, and I announce that you shall be crowned at Rheims." She added, after a pause, "Why will you not believe me? God has pity on you and on your people; for *St. Louis and Charlemagne* are on their knees before him, praying for you and them."

Her prophetic power was very clear. A soldier coarsely jested her in the streets, when she quietly replied that it did not become a man so near his end. That same day he was drowned. When equipped for war, in knight's armor, she declined accepting a sword, saying that there was one with five crosses lying in the church vault of St. Catharine's, and this, and none other, she would have. A messenger was sent, and the old, neglected sword found, as she had predicted. A banner was made, as she directed, and the assembled army saw her with exultation. Thousands of deserters again enlisted; so wild was the army's enthusiasm that she at once became their chief.

She said that in seven days she would raise the

siege of Orleans, and on the seventh day the English departed.

She said she would go out of Orleans in the morning, and return by a bridge then occupied by the enemy, and she accomplished the superhuman feat.

She, after the siege of Orleans was raised, desired the king to go to Rheims and be crowned. While he debated in his mind whether it would be expedient to ask her what her spirit said on that subject, she read his thoughts, and exclaimed, "You desire to know what the voice says. I heard it declare, Daughter! go forward; I will be thy helper — go! And when I hear that voice, I feel so joyous, it is too wonderful to tell."

The expedition to Rheims was full of danger, for the place was still in the hands of the enemy; but the king concluded to go. He first besieged Troyes, but failing in provisions, a council of war recommended a retreat. It was, however, interrupted by Joan, who exclaimed, "The city is yours if you remain before it two days longer." She now fulfilled her prediction, and without encountering a single shot he marched his army into Rheims, and was crowned King of France.

At her trial for witchcraft by the English, she predicted that Paris would be lost by them within seven years, which was also verified.

Lastly, Christ-like, she foresaw her own destruction. "I shall only continue for a year or a little more," said she. "I must try to employ that year well." Her character sets scepticism at defiance, and on no other supposition but the interference of a superior intelligence can we account for the sudden transformation of a retired, modest shepherd girl into a hero possessing the original fire of the prophets of old.

Of greater influence were the beams of spiritual knowledge the spirit world poured through the mediumistic powers of the sages and prophets of old: Menu, Zoroaster, Confucius, Christ, Mohammed, and Swedenborg, and an opposite influence which impelled the conquerors of the world to scourge mankind, and an Alexander and Napoleon to become more destructive than the most savage beasts.

4. *Apparitions.*— By a gradual ascent we are led to apparitions, in which hallucination becomes, as it were, intensified, and assumes definite form. Mediums see spirits not by their natural, but by their spiritual eyes. Before the contrary is asserted, let us reason. Spirits cannot be seen by the mortal eye. The substance of which they are composed does not reflect light, and hence cannot be seen. We well know that the contrary is maintained, but very erroneously; for if you assert that you have seen a spirit in your normal state you assert a contradiction. You were either unconsciously clairvoyant, or you did not see a spirit. We can collect particles, as of phosphorus, and produce a visible image; but this image is not ourselves, and all such manifestations should be regarded with distrust until thoroughly examined.

When a medium sees a spirit, he sees it just as the mesmeric subject sees objects he is willed to see by his magnetizer. He sees images formed in the mind of the controlling spirit. This subject will be extended, and the proposition proved hereafter.

Grief often develops susceptibility, and mourners often receive consolation in the hour of trial which they are incapable ever after of receiving. We will introduce a pleasing anecdote illustrating this, as well as the retention of character by spirits.

A gentleman, who had recently lost a dearly beloved wife, was, in company with his son, crossing a lonely moor several miles from his home, when they both saw the wife and mother seated on a mossy stone by the side of the path, and the faithful house dog lay at her feet. She soon after vanished, but the dog remained; and as they approached, expecting every moment that he, too, would disappear, they found, to their surprise, that it was the veritable house dog of flesh and blood that met them. The apparition was produced by the wife's seizing the favorable moment presented by their really seeing the dog on the moor. They could not conceive what could have taken the dog there, but animals are often highly susceptible, and he was urged by her will.

This well-authenticated fact shows in a very striking manner the peculiarities of spirit life. By a keen instinct, or mesmeric susceptibility, the dog felt the presence of his beloved mistress, and followed her unseen shadow; she, retaining her affection for the faithful animal, and love for her accustomed walk on the moor, still rambled through the familiar scenes.

Such facts lead us to the consideration of presentiments which in chameleon forms present themselves so inexplicably that the learned have, in order to veil their ignorance, referred the whole subject, with its host of facts, to hallucination, or rather to rank folly. To show when or how we account for this class of facts on scientific grounds, and where we draw the line of demarcation between facts referable to spirits, and those of a mundane origin, we introduce a short extract, which, from its condensation, forbids further concentration.

"Among all the branches of the supernatural, there is no one which has been so little discussed by philosophical writers as that known by the term *presentiments.* And yet there is no one among them all better entitled to our consideration, from the many and well-authenticated instances which may be cited to prove their existence; nor is there any one of them, at the same time, so difficult of explanation, on natural principles, when that existence is established. It is this difficulty, probably, which has deterred many learned men from attempting a solution of the mystery, while it is the secret reason, we apprehend, why many others pass the subject with a slur, placing the presage to the account of despondency of mind, or nervous timidity, and professing to look upon its fulfilment as nothing more than one of those remarkable coincidences which are often occurring in the ordinary events of life. This is doubtless an easy way of getting along with what we will not believe, and cannot explain; but it so happens that by far the greatest proportion of the recorded cases of presentiments (by which term we mean forebodings which are realized, not false presentiments) has occurred among a class of men the most noted for firmness and courage, the least subject, by nature and discipline, to be affected by superstitious fears or nervous weakness. Scarcely an important battle has been reported, by the details of which it has not appeared that some of the slain, though the bravest of the brave, and never before troubled with such impressions, have foretold the death that awaited them.

" It was once our fortune to be thrown into a social circle, in which were the relatives of some of those who perished in the conflagration of the Richmond Theatre, in 1812, which so widely scattered the weeds of woe among the first families of Virginia. Two or three remarkable instances of presentiments were told us as having been felt and avowed previous to the fire by those who became victims ; but we have treasured up one more peculiar than the others, because, instead of being followed by the death of him who was the subject of the premonition, it was the direct means, in all human probability, of saving him and a family of accomplished daughters from destruction. The play announced for the night was an attractive one. The gentleman to whom we allude had proposed to his family to attend the theatre with them, and several times through the day spoke of the pleasure he anticipated in witnessing the performance. But towards night he became unusually thoughtful, and, as the appointed hour drew near, he took a seat with the ladies, and commenced reading to them a long and interesting story, evading all conversation about the theatre. This he continued until interrupted by one of the wondering circle, who suggested that it was time to start. Again evading the subject, he went on reading till he was a second time interrupted, and told they must go immediately or they should certainly be too late. Finding he could not put them off till too late to go, as he had hoped to do, he turned to them, and earnestly asked it as a favor that they would all forego the promised pleasure of the playhouse, and remain with him at home through the evening. Though deeply surprised, and sorely disappointed, yet they dutifully acquiesced; and, in the course of the evening, while engaged in their quiet fireside entertainment, they were aroused by an alarm of fire, and in a few minutes more by the appalling tidings that hundreds were perishing in the flames of the burning theatre, in which, but for the request which had seemed so strange to them, they, too, would have been found to be numbered among the victims. The next morning the gentleman told them, in explanation of his conduct the evening before, that, as the hour set for the performance approached, he became unaccountably impressed with the idea or feeling that some fearful calamity was that night to fall on the company at the theatre; and that the premonition, in spite of all his efforts to shake it off, at length became so strong and definite that he secretly resolved to prevent them from attending, and would have done so even to guarding the door of his house with loaded pistols.

"One more instance must we relate in illustration of our subject, which is that of an adventure which was once related to us by an intelligent, truthful, and highly-valued friend, and which we will give in his own words. 'Some years ago,' he said, 'I was journeying on horseback through a part of the wild and sparsely settled country lying west of the Mississippi, with about two hundred dollars in silver and gold stowed away in my saddle bags. After having travelled one afternoon till nearly sunset, without seeing a single hut or inhabitant, and while anxiously casting about for some shelter for the night, I had the good luck, as I esteemed it, to overtake a very honest-looking squatter, of whom I inquired the distance to a tavern. He said it was fifteen or twenty miles, quite too far for me to think of going that night, but if I would go with him to his cabin, which was a mile or so off the road, I should be welcome to such accommodations as he and his wife could furnish me. Being taken by the plausible and apparently kind manner of the man, I thankfully accepted his offer, accompanied him to his log hut, and was hospitably provided with refreshments. When I retired to my bed, which was on the lower floor, and adjoining the room occupied by my entertainers, with my saddle bags, which I had unwisely let the man handle, placed under part of my pillow, I soon fell asleep with feelings of the utmost security, having no suspicion that my entertainers were not kind and worthy people. After sleeping a while I awoke restless and uneasy—why I knew not; I thought I must be sick, and fell to examining my pulse, &c., but could detect in myself no symptoms of illness. Besides, I soon found that my uneasiness was not like that of any physical illness. It was a feeling of apprehension—a vague, yet strong impression that some great evil or danger was impending over me. I tried to reason myself out of such folly; but, instead of succeeding, soon found the strange feelings growing too intense to permit me to keep in bed any longer. And accordingly I arose, crept stealthily to the door opening into the other room, and listened. I could soon distinguish the voice of the man and his wife, who seemed to be engaged in a low and somewhat flurried conversation, of which I at length caught enough to convince me that they were planning my death, and the manner of disposing of my body afterwards. I hastily crept back, dressed myself, and drawing out my pistols, awaited the result.

Presently the door opened, and I caught a glimpse of the man entering with an axe in his hand, and approaching on tiptoe towards me. Instantly cocking my pistols, I called on him to stop, or I would shoot him dead on the spot. He was evidently taken by surprise; for, turning about with the quickness of thought, he hastily skulked out of the room. After watching, with my pistols in my hands, till the first appearance of daylight, I made my escape, unheard, from the house, mounted my horse, and departed with all possible speed. Gaining the road, I rode on, and in about five miles, instead of fifteen, came to a tavern, where I ascertained that the man at whose house I had staid was strongly suspected of having decoyed several other travellers to his cabin, in the manner he had me, and murdered them for their money.' "

The gentleman who was impressed with the burning of the theatre received what is called a remarkable presentiment from his spirit guardians. The other gentleman, who was impressed with the murderous design of his host, probably received it from a similar source; but we can as readily account for the fact, and probably in a more philosophical manner, by considering him to have been very impressible, and the intense thought of the host, directed especially to him, aroused and warned him of danger.

Of purely spirit impression Swedenborg presents an historic and renowned illustration. Mrs. Crowe has, in her Night Side of Nature, collected a great variety of kindred facts. We select these facts, showing how impressions can be made, under favorable conditions, so clear and distinct as to appear to the recipient as audible sounds, or a clear, articulate voice.

" Grotius relates that when M. de Saumaise was councillor of the parliament at Dijon, a person, who knew not a word of Greek, brought him a paper on which was written some words in that language, but not in the character. He said that a voice had uttered them to him in the night, and that he had written them down, imitating the sound as well as he could. M. de Saumaise made out that the signification of the words was, 'Begone! do you not see that death impends?' Without comprehending what danger was predicted, the person obeyed the mandate and departed. On that night the house that he had been lodging in fell to the ground.

" An American clergyman told me that an old woman, with whom he was acquainted, who had two sons, heard a voice say to her in the night, 'John's dead!' This was her eldest son. Shortly afterwards, the news of his death arriving, she said to the person communicating the intelligence to her, 'If John's dead, then I know that David is dead, too, for the same voice has since told me so;' and the event proved that she was correct.

" A Mr. J. related a singular personal experience to Mrs. Crowe. He had been ill, and there being no apothecary in the immediate neighborhood, had been accustomed to send to a village some five miles distant to procure medicine. One night he had been to M. for this purpose, and had obtained his last supply, — for he was now recovered, — when a voice seemed to warn him that some great danger was impending — his life was in jeopardy; then he heard, but not with his outward ear, a beautiful prayer. 'It was not myself that prayed,' he said; 'the prayer was far beyond any thing I am capable of composing. It spoke of me in the third person, always as *he;* and supplicated that, for the sake of my widowed mother, this calamity might be averted.' It appears, from the further details of this case, that, when Mr. J. was about to take his medicine, he fancied there was something peculiar in its appearance, and his suspicions were excited. He hesitated, but at last took half the prescribed quantity. This, however, was speedily followed by the most alarming symptoms; the chemist had made a mistake; the compound contained a deadly poison, and, notwithstanding the smallness of the dose, the patient with difficulty survived its effects."

Jung Stilling was a highly susceptible medium. When writing to his friend Hess, he experienced a deep interior sensation, as though a voice spoke to him, saying, "Lavater will experience a bloody death." Two months after Lavater was shot by a Swiss grenadier.

Captain Griffith retired while at sea, presuming he was several hundred miles from land. He had scarcely closed his eyes in sleep when he heard the cry of "Breakers ahead!" He started up, ran on deck, but perceiving no danger he again retired. He was aroused again by the same cry, ran on deck as before to find all safe. He was aroused the third time by a light under the lee, the same terrible cry, mingled now with the hoarse sound of the breakers. He again rushed on deck, and now found his presentiment true; he was rapidly approaching the shore, on which a violent sea was breaking, and narrowly escaped wreck by vigorous exertion.

A similar fact may be philosophically referred to

spirit intercourse. A widow and her son dwelt together after the decease of her husband. The son was very vicious, and maltreated his mother. One day he heard the voice of his spirit sire: "I have seen your treatment of your mother. Go and do better hereafter, or I shall appear to you."

Another incident is related of a clergyman of the north of England quite as remarkable, and only referable to interposition of spirit guardians.

"It was winter, and a severe snow storm prevailed at the time. He was pursuing an unfrequented road, which was obscured by the heavy fall of snow. Evening came on, and the deepening gloom rendered it impossible to determine whether he was riding in the right direction. However, he continued to wander on, though unable to perceive any sign of human habitation, and doubtful whether he was every moment drawing nearer to his destination or to destruction. At length night invested the dreary landscape, and all outward forms, in her soft mantle woven of the shadows, and the traveller began to realize more deeply the nature of his situation. He felt some apprehension, and his fears struggled with his confidence in the divine Providence, when, suddenly, his meditations were interrupted by a loud voice, that seemed to come from the upper air, with the startling power of a trumpet blast. The voice uttered, as nearly as I can remember, the following emphatic words: 'Stop! Stop! Stop! Turn about! Turn about! Turn about!' The horse stood still, and his rider, instinctively obeying the voice, turned the animal round, when he perceived, a little off from the direction he had come, a light that seemed to indicate the locality of a dwelling. Instantly inspired with the hope of finding a place of security from the dangers of the night, he directed his steps towards the light, and soon found that it shone from the window of a cottage, where he obtained a comfortable shelter. The storm subsided about the same hour, and on the following morning, the tracks of the horse being distinctly visible, he felt a curiosity to visit the spot where he was arrested by the mysterious voice. Accordingly, he pursued the path to its termination, and was utterly amazed to find himself standing on the very brink of a chalk cliff some two hundred feet above the water! Had he proceeded ten feet farther he would have plunged into the abyss below!"

The horse heard the voice as well as his rider; and many facts might be produced showing that animals are very susceptible to magnetic and spiritual influence — a fact acknowledged thousands of years ago, when Balaam's ass saw the angel blocking up his way.

We introduce two facts proving that spirits can foretell future events, especially the time of the death of individuals.

Mr. R. conceived the idea that he was going to die, and immediately began preparing for the change. In three days, or by the time his business relations were settled, he was a corpse.

Mr. G., while in perfect health, was impressed that he was shortly to die. He went to the undertaker's, engaged a coffin, returned home, was shortly after taken sick, and at an early hour in the evening expired.

Such facts might be multiplied ad infinitum; but

of what avail to relate the ghost stories of the ages? We are not here endeavoring to establish a system, but to work out the path of rational investigation; on the one hand, to reduce certain facts to spiritual agency, from the ranks of sceptical philosophy; on the other, to account for many supposed spiritual manifestations by mundane causes. Do not say spiritualism will suffer by such a course. The truth can never be injured, and sooner or later spiritual science must be classified, and the power of spirit, while in the body, be fully recognized. It is best, the part of policy, even, to place all facts, each on its own basis, and not for the sake of puerile argument force a distorted meaning from any.

Volumes of such facts might be gleaned from history, but this must suffice to represent its class; for we are less ambitious of increasing the size of our book than condensing our ideas.

The ghosts of delirium are often creations of the brain; its thoughts, clothing themselves with reality, or being incarnated, appear as real personalities; but often the moment of excitement is seized by depraved spirits to haunt the suffering mortal.

It is a rash evasion of this question to say that all the apparitions recorded in the world's history are delusions. The facts exist, and cannot be evaded or explained away. One billion of incarnated ghosts dwell on earth to-day, among which billions of airy beings stalk unseen. There are ghosts enough. Often do they possess unlimited power.

This subject is naturally divisible into two classes of phenomena, wholly different in their cause. By not recognizing this division, but referring all apparitions to a common cause, philosophers have signally failed in their explanations.

One class may be considered of mundane origin, taking their rise in the mind by derangement — mental or physical. These are disordered, meaningless, and confused. The other class, which are spiritual in their origin, are embodied spiritual impressions, and are usually significant, generally prophetic, as are dreams of like origin. And the wise will obey monitions thus received; for nothing but a great emergency ever calls the interference of superior beings. Remember, man is thrown on his own resources; and we have neither inclination nor power to interfere with his individuality. He must lay his own plans and abide the consequences; but if danger threatens, we can, prompted by our love for our friends, give timely warning, which, if heeded, will save them from the consequences of blind efforts.

5. *Dreams.* — The physiologist says, "We can no more account for dreams than for thoughts;" but we think we have, at least partially, accounted for thought, and dreams, by the spiritual philosophy.

Sages have failed to account for dreams, because they sought a common cause; whereas two distinct causes may be mentioned, wholly unlike, and hence confusion and uncertainty have pervaded all investigations and theories on this subject. These causes divide dreams into two natural classes: —

(1.) Dreams resulting from psychical or physical derangement.

(2.) Dreams which are spiritual impressions received in partial sleep, or a state approaching unconsciousness.

(1.) *Dreams resulting from psychical or physical derangement.*

All that has been written applies to the first class, and hence wholly fails to account for the phenomena of the second, which are of a far more practical and 'nteresting character. A writer, speaking of dreams, remarks, —

" We form our dreams by referring any idea that occurs to some class of thoughts which had before passed through our mind. Thus a person, who had fallen asleep with his face towards a narrow stream of light, immediately began to dream that a column of darkness grew up before him. The idea of this darkness would, we know, be excited by the eye being directed towards the light. Speedily this black column began whirling along over a wide plain. This idea of motion was probably excited by motion of the eye; but it was no sooner perceived or imagined, than the mind began to explain it, by associating it with what had been heard concerning columns of sand carried before whirlwinds. Immediately he seemed to be in the burning deserts of Africa, with the red sun on the verge of the horizon, while the vast column of sand was hurrying to overwhelm him; but in a moment some miracle saved him, and he awoke. Now, it is evident that physical phenomena produced the sensations which excited the mind; but the mind itself made the dream, partly of memory, partly of sensation. Then, again, the manner in which the mind goes back to the past for its ideas in dreaming, suggests the profundity of mystery which belongs to the subject, and at the same time informs us that the operations of the mind are not to be explained by the anatomist. Why did Huber, after forty years of total blindness, dream of the sights familiar to childhood? If dreams result from reflex action of the brain, and the images conveyed through the senses are reproduced only because the nerves physically retain their impressions, then have we the vast marvel of material substances preserving in themselves ten thousand pictures of the past, all mixed together, yet not confounded, each dependent on a particular state of the nerves; and yet all particular states existing at once in a latent state, and every image of the countless multitude fixed in the nerve — matter capable of being spontaneously represented and recognized by that matter. Dreaming and delirium are but memory modified by the state of the will in relation to the body. Hence, aged persons are apt, in mental absence, — whether asleep or awake, — to behold the scenes familiar to youth, and in imagination so to associate with the. dead as sometimes not to be able to distinguish them from the living. It is no uncommon thing for such persons to sleep soundly, and yet say they have not slept at all; and that merely because their dreams are so vivid and distinct that they confound them with realities; and in that kind of delirium, frequently experienced in the feebleness of old age, the features, the dress, the language of friends, are exactly recalled, after scores of years passed in apparently entire forgetfulness of them."

To the spirit in the dream-state, time and space are annihilated. Like a real spirit, the dreamer desires to be in such a place, and thinks he is there.

In a state of perfect sleep dreams never occur, because all faculties, perceptions, and thoughts are, as it were, dead; so that the supposition that the mind always dreams, but fails to remember its wandering fancies, is untrue. When some faculties remain awake, from over-exercise or a deficiency of exertion, while others sleep, dreams result. The same takes place when the body has been over-worked, or not exercised sufficiently.

A person partakes of salt food for supper, and retires thirsty. Immediately he dreams of running streams and fountains, which he cannot taste, though dying with thirst. Such dreams usually occur in fevers. A traveller in a remote land having the ague, when the fever was on, could, even when awake, see nothing but the icy fountains gushing out from the sides of his native hills.

A hearty supper often disturbs the harmony of the system by producing a fever, and originates terrible dreams; or the pressure of the stomach on the great contiguous veins, interrupts the circulation, and dreaming gives place to nightmare; when suddenly disturbed while sleeping, in the moment between sleep and consciousness, confused thoughts array themselves in form; and it seems we have dreamed for hours, when perhaps all passed in a second through the mind.

(2.) *Dreams produced by Spiritual Impressions.*

Having thus rapidly glanced at the origin of one great class of dreams, we approach the other and by far more interesting division. We have shown what is not, we will now show what is, spiritual.

A person dreams of travelling to a certain country, and sees the scenery as perfectly as though physically present. Years after he really travels there, and is astonished at the familiar aspect of the scenery. He recollects his dream, and finds it true to the letter. Such dreams can be referred either to clairvoyance or to direct spiritual communication. Clairvoyance may counterfeit normal sleep; or, in other words, the person really thinks he sleeps, when he is really clairvoyant. Many persons are highly impressible while asleep who are not so when awake. Hence the hours of slumber are employed by guardian angels to impress ideas beneficial to the recipient. These take the form of dreams, and are generally prophetic. Some imminent danger calls them forth, and they should always be heeded.

If prophetic dreams thus originate, why does not the spirit impress at once the real things, and not, as is universally known, speak enigmatically, and by symbols? This objection equally well applies to the prophetic oracles of all ages, and its explanation is easy.

Suppose some great danger hovers over the sleeper, and his guardian spirit desires to warn him of it: if he should impress the real danger, he would startle at the first sentence, his mind become excited, the necessary conditions of receptivity would be destroyed, and it would be impossible to proceed. On the contrary, when symbols are employed, the mind remains passive, not knowing what is to come, or the meaning of that which it has already received, until all is given; and in waking moments it reflects on the meaning of these symbols, which usually are sufficiently clear to allow of their meaning being readily obtained.

Laugh at the fantasies of a fevered brain, or the visions produced by a gorged stomach — the nightmare of the gormand, and ghost-seeing of the dyspeptic; but the dreams of the clear head and pure heart are words of angel visitants, and should be treasured and observed. When man rests in the arms of Sleep, she hushes him by the hymns of angelic voices and sweet visions of the morning land.

6. *Influence of Mediums.*— We shall hereafter allude to the influence mediums exert on our communications. As is the channel, so the stream which flows through it; the vessel gives form to the water which it contains. This influence is like that which is exerted on a psychometrist by an autograph, or on a magnetic subject when he endeavors to read the thoughts of others.

The character of the medium, in a great measure, determines the character of the spirits who control him. There cannot be any very great difference of development between the controlling spirit and the medium. There must be affinity. Each medium attracts a class or grade of spirits peculiar to himself. These are attracted: first, because certain phenomena can be given through his organization which cannot through any other; second, because certain thoughts can be transmitted through his brain, which cannot through any other; third, because there is similarity or congeniality.

Some mediums are used entirely for physical manifestations, and never for communication. They attract spirits capable of producing such phenomena, who act entirely by means of the organic peculiarity which constitutes such mediums. Others are entirely used for writing or speaking. Such are impressible, and attract generally a higher grade of spirits The intelligence of the communicating spirit is in direct relation to the intelligence of the medium. Mediums of a scientific and philosophical cast of mind attract wise and sagacious spirits, willing and capable to instruct on those subjects; whereas minds of a poetic temperament attract poets, and generally write poetry. A grovelling, low-minded medium attracts spirits from the lowest stratum of the spirit world, and receives according communications. The state of the medium, while receiving communications, also determines their character. When exhilarated by the flow of health, happy and cheerful, developed beings can enter into the chambers of the soul, and breathe forth beautiful thoughts; but when the nerves are wasting by disease, and the system jars with inharmonious vibrations, undeveloped beings enter the door thus thrown open, and poison the springs of thought.

Mediums and spiritualists should ever keep these facts in mind. They should know that they are constantly surrounded by all grades of intelligences; and wherever a channel of communication is opened, they enter in. We need not impress the necessity of living pure and elevated lives in order to attract the purest angels of the spirit land. We love the spotless soul, and hover around the undefiled in heart.

7. *Influence of Conditions.* — There are essential conditions which must be fulfilled for us to communicate. If by impression, we must have an impressible medium; and it is useless to add that the correctness of our communications is directly as the impressibility of the medium. If but partially impressible, his thoughts will mingle with ours, until our meaning is

lost, or distorted into something entirely different, as is often the case.

The circle, too, has an influence on the mind of the medium, and may compel him, when our influence is feeble, to reproduce their own thoughts and desires. The same is true of surrounding persons not in the circle. A medium is necessarily extremely susceptible—as susceptible as the needle trembling to the pole, and quivering to the slightest disturbing cause; and hence, unless cautious and guarded, is liable to be imposed on.

The electrical state of the atmosphere is often a source of failure, when it is antagonistic to our influence. So the sphere of persons who are discordant repels elevated and attracts low spirits.

8. *Influence of the Communicating Spirits.* — We need not remark that the grade of development of the communicating spirit, other things being equal, determines the degree of intelligence of the communication. But the communications of all must not be received as truth, for falsehood abides here, and is ever ready to thrust itself upon the notice, and mislead its dupes.

The sources of fallacy in a spirit communication are, 1. It may become mixed, or entirely perverted, by the mind of the medium. 2. The medium may be influenced partially by the circle; and hence the communication partakes of their thoughts more than of the spirit's. 3. Low and depraved spirits may assume well-known names, and communicate vague or erroneous ideas. 4. Well-meaning but ignorant spirits may communicate ideas they sincerely believe, but which are nevertheless false.

Such are the sources of error, which the intelligent investigator will always strive to avoid.

Thus we have endeavored to draw a line between closely related phenomena, and show what must be referred to spirits, what to other resources, and present a rational system, not to supersede the thoughts of the reader, but to compel him to think the more.

CHAPTER VI.

SPACE ETHER.

Space. — Incomprehensibility of Distance. — Of Minuteness. — "Air of Heaven." — Conjectures of the Indian Philosophers. — Of the Ionian School. — Of Pythagoras. — Of Empedocles. — Modern Speculation and Demonstration. — Olbers's Proof. — Herschel's Statement. — Limited Transparency of Space. — Retardation of Comets. — Planetary Motions. — Of Space Ether.

THE suns and worlds of the universe float in an infinite ocean which we term *space*. Philosophers in all ages have indulged in speculations and dreams as to its nature. Some have doubted its infinite extension, as Aristotle.* None, however, doubt its immeasurability; for figures illy express the distances the smallest telescope reveals, and the array of figures denoting the distance of the nearest stars is mere puerile pedantry. A million of miles is as incomprehensible as a hundred millions. All our ideas are comparative; and when numbers exceed our objects of comparison we utterly fail to grasp their meaning. What idea can we form of the distance of the star 61 Cygni, when told that it is 62,415,000,000,000, or sixty-two trillion four hundred and fifteen billions of miles from us? or of the remoteness of the Milky Way, when Herschel † tells that us his twenty feet reflector revealed stars from which a ray of light, departing when Christ was crucified, and travelling

* Aristotle, De Cœlo 1, 7, p. 276.
† Outlines of Astronomy, § 803, p. 541.

two hundred thousand miles per second, would scarce reach us in the year 2000. All attempts to grasp such numerical statements must utterly fail. Equally incomprehensible is the minuteness of organic life. The number of organic beings in a cubic inch of chalk is about equal to the number of miles to the nearest star. In ten cubic inches of Bilin polishing slate are animalcula enough to distribute one for every mile from our earth to Sirius. Humboldt well observes that such estimates remind us of the treatise of Archimedes, named the *Arenarius*, in which he computes the number of sand grains which might fill the universe of space. Had Archimedes viewed a grain of polishing slate with a microscope, he, perhaps, might have appreciated the inapplicability of numbers, when he realized the startling fact, that to write out the number of beings in his knapsack full of slate, his line of figures would encircle the earth.

Although it were fruitless to conjecture the limits of space, the investigation of its properties falling within the limits of observation is more productive. The propagation of light, in its special form, and the seeming action of a resisting medium on Encke's comet, and also the evaporation of the tails of some large comets, seem to prove that space is not a void, but filled with some kind of cosmical matter. This has received the names of "air of heaven," "cosmical matter," and "space ether," significant of theories entertained by their originators. The idea of an all-pervading ether is very ancient. The Indian philosophers made it one of the five elements, and described it as having infinite subtilty, and as the medium of animal life. Its name (*âkâ'sa*) denotes that they also regarded it as connected with light.* The Ionian philosophers considered this ether wholly distinct from the atmosphere which surrounds the earth. It was "of a fiery nature, a brightly beaming, pure fire-air, of great subtilty, and eternal serenity." † They probably identified it with fire, as Anaxagoras believed that the upper regions are full of fire, in incessant motion; and by a play of imagination peculiar to the Greeks, it became an incarnation of the Divine, unlike any thing else. ‡ The rarification of air on high mountains, its purity increasing with the height, seems not to have suggested the idea of pure ether. The Greeks were not accustomed to, nor organized for, observation. Their philosophy was purely speculative, and they were guided more by intuition than phenomena. While Aristotle placed the ether among the conditions of matter, Pythagoras made it a fifth element, represented by the fifth of the twelve regular solids. §

The purely imaginary ether of Empedocles in modern times becomes demonstrated; and physics refer the transmission of light, with all its collateral phenomena, to its agency. The ether, as described in the Aristotelian philosophy, penetrates all living organisms, and in them becomes the principle of vital heat—the very germ of a physical principle stimulating man to independent action. The main difference between the ether of moderns and ancients is, that, while the latter considered it brilliantly luminous,—as seen in certain

* Referred to by Strabo, according to Megasthenes, xv. p. 713.

† Anaxagoras and Empedocles.

‡ Plato, Cratyl. 410 B.; Arist. De Cœlo, 1, 3, p. 270; and Arist. Meteorolo. 1, 3, p. 339.

§ Martin, t. ii. pp. 245–250.

phenomena through rents and chasms in the solid sphere of the firmament, — the former consider it the medium of transmission of luminosity. While the old philosophers idly speculated on its existence, and built castles in Dreamland, it was reserved for modern research, both in mechanics and mathematics, to demonstrate the existence of this all-pervading medium. There are two observations which substantiate its existence — the limited transparency of space, and the retardation of comets. Eighty years before Olbers of Bremen, Loys de Cheseaux called attention to the fact that if space was perfectly transparent, as we could not conceive the smallest point in the firmament destitute of a star, the whole heavens should glow with the radiance of the sun's disk. Otherwise we must consider that light follows other laws than are given it, and decreases in a greater ratio than the square of the distance. Such are the reasons which convinced Olbers, Cheseaux, and Struve, that space is not absolutely transparent.* Herschel admits that if the light of Sirius lost one eight-hundredth of its intensity in passing from that star to the earth, which gives the amount of density of a fluid capable of diminishing light, it would suffice to explain the phenomena. But he has cast doubts on this evidence, and his investigations seem partly in a different direction, although not in reality; for if he saw stars projected on a black ground, he only looked through our star-stratum, and the distance to the next, possibly may be beyond the power of telescope, the

* Olbers, in Bode's *Jahrbuch*, 1826, § 120, 121. Struve, Etude de Astr. Stellaria, 1847, p. 83–93. Outlines of Astronomy

light emanating from such remote creations being too feeble to produce even the dimmest luminosity.

The retardation of comets affords more positive testimony. M. Encke found that his comet was accelerated about two days at each revolution, and that this occurred as though produced by a resisting medium. Contradictory as it may appear that resistance will shorten the periodical time of a comet, it, nevertheless, depends on planetary laws. The planets are held in their orbits by two forces — that which chains them to the central orb, and that which tends to make them fly off in a straight line. If one of these forces is diminished, the other will increase in exactly the same ratio. Thus, if this medium offered resistance to a comet, it would fall towards the sun, and being acted on more strongly by gravity it would move not only in a smaller orbit, but faster in that orbit. The other comet, which returns after a period of six years and three quarters of a year, has been accelerated a whole day in its last revolution. This places the existence of cosmical ether beyond a doubt, for if it exists it must be subject to the laws of gravitation, and hence be more dense around the sun than in the free region of space — a presumption proved by this comet, for while it revolves nearly between the orbits of Earth and Jupiter it is only accelerated one day. Encke's, revolving between Mercury and Pallas, is accelerated two, showing the increase of density towards the sun; which is still further shown by the increase of resistance during the twenty-five days the comet is nearest that luminary. This fluid cannot rest, at least when brought in contact with a system like our solar system, which moves, but must rotate, in common with it around a great vortex. Hence its action must be much more strongly felt by comets that move in a conflicting direction than those which move with it — an inference supported by observation. Such is a primary idea of the all-pervading ether, to which, as effects, the so-called imponderable elements are to be referred.

To these agents we now turn, and shall examine them one by one, pointing out the differences and similarities which exist between them.

CHAPTER VII

PHILOSOPHY OF THE IMPONDERABLE AGENTS, AND THEIR RELATION TO SPIRIT.

LIGHT. — Its Velocity. — Bacon's Conjecture verified. — Analysis of. — Philosophy of Colors. — The undulatory Theory. — Length of Waves. — Collision of Waves. — Newtonian Hypothesis. — Proofs of the Wave Theory. — Arguments against the Theory of Transmission. — Objections considered.
HEAT. — Analysis of Solar. — Its Relations to Light. — Referable to a common Cause.

LIGHT emanating from all luminous bodies is propagated with uniform velocity. The observations made on the satellites of Jupiter, and the sun, coincide perfectly with those made on the earth. This fact accords with the hypothesis of a universal elastic medium, and proves that its elasticity is precisely equal to its density. The velocity of light is one hundred and ninety-two thousand miles per second — a motion so rapid that the mind can scarcely grasp it without an illustration. Voyagers are usually three years circumnavigating the globe, but a ray of light would circulate around it *nine times* during the single tick of a clock. Great discoveries are always preceded by conjecture. Bacon entertained the idea that light was not propagated instantaneously, "for it seems incredible that the rays of celestial bodies can pass through the immense intervals between them and us in an instant, or that they do not require some considerable portion of time." He scarcely could hope for so early a realization of his conjecture, or that instruments would be constructed so perfectly, as to catch the dim light of stars so remote, that, if blotted from existence, ten thousand years would intervene before the last ray which departed from them reached us. Newton and his immediate followers supposed that light was a material substance emitted by luminous bodies, and flowed in straight lines with prodigious velocity. The impinging of these on the optic nerves produces the sensation of sight. This hypothesis fails to account for a tithe of the facts observed, and is untenable. In order to arrive understandingly at a true theory of light we must examine its properties.

White light is a compound of four different colors combined in certain proportions. The method of analysis is to admit a small ray into a darkened room, and then make this ray pass through a prism or wedge of glass. According to its refrangibility (length of wave) each color will arrange itself in the spectrum, and instead of a colorless spot, a beautiful ribbon, colored like the rainbow, will appear. The order of the colors, commencing with the rays of greatest refrangibility, is the same as seen in the rainbow — lavender, violet, indigo, blue, green, yellow, orange, red. The red, yellow, blue, and lavender, are the primary colors, the others resulting from admixtures of these.

The color of natural bodies results from the destruction, or, as it is commonly expressed, absorption, of all other colors but the one or ones reflected.

With light thus analyzed, or reduced to its simplest elements, experiments can be performed, proving, with almost mathematical certainty, that they are vibrations in an ethereal medium, and not emanations.

When two equal rays of red light fall on the same

point, they produce a red spot twice as bright as either, providing their lengths be exactly 0.0000.258 of an inch; or if their length be twice, three times, &c., that distance. But if the length of the two rays be one half, one and a half, &c., that distance, one ray will extinguish the other, and total darkness result. If the two rays be $1\frac{1}{4}$, $2\frac{1}{4}$ of the 0.0000.258th of an inch, the red light of the combined beams will be equal only to what one alone would produce. Violet light may be used, but the length of the two rays must not be equal, but differ by the 0.0000.157th of an inch, and for the other colors each has its own difference. The same result is obtained by viewing the flame of a candle through two very fine slits, extremely near to each other, in a card, or by admitting the sun's rays into a darkened room, through a pin-hole 1-40th inch in diameter, receiving the light thus admitted on a sheet of white paper, and holding a slender wire in the light. The shadow of the wire will be extremely light in the centre, with a brilliant series of colored stripes on each side. The waves which bend around the sides of the wire are of equal length, and hence produce a double brightness in the centre; but the waves which fall outside of the centre are of unequal lengths, and destroy each other, producing black lines. When a pure prismatic ray is used, as red or blue, only black lines and the given color are seen; but when white light is employed, from its mixed nature, the black lines alternate with fragmentary colors.

That this phenomenon actually results from the waves flowing round the wire, is proved by interrupting those on one side. The colors vanish instantaneously. We cannot suppose that two particles of matter annihilate each other, but we can readily suppose two opposing motions can produce an equilibrium. If two equal waves of water meet, they mutually destroy each other. So of waves of air; and this phenomenon of light is precisely similar. If two waves moving in the same direction with equal velocity unite, the resulting wave is equal to their sum; so two waves of light uniting produce a double brilliancy. These waves are excited in the light-ether by all luminous bodies. It is supposed that the particles of light-emitting bodies are in a state of constant agitation, and they excite waves in the ether corresponding to their own vibrations, which are transmitted from particle to particle, the particles remaining at rest, while the motion is transmitted like a wave in water. Although the progressive motion of light is in a straight line, the vibration of its particles has been found to be at right angles with the plane of direction. For illustration, a ray of light may be compared to a long rope through which vibrations may be sent, yet itself remain at rest.

The intensity of light depends on the extent of the vibrations, their length and frequency. The length of wave and time of vibration have been very ingeniously computed. As the velocity of light is known, if the length of the waves is desired, the time corresponding to each is easily determined. All transparent substances of a certain thickness and having parallel sides will transmit and refract light; but if made extremely thin, the transmitted and reflected light is colored. The vivid iridescence of soap bubbles and many minerals is thus explained. If a plate of glass be laid on a lens of almost inappreciable convexity, a black spot will appear at the point of contact, surrounded by seven colored rings. That these rings are formed between the two surfaces in contact is proved by laying a prism on the lens instead of the glass plate, so that reflected light is cut off, and the spaces between the different colors appear perfectly black — a strong proof of the undulatory theory, and one considered conclusive by all scientific men. According to this theory these spaces ought to be perfectly black, whereas, according to the Newtonian hypothesis, they should be partly illuminated. If, instead of a plate of glass, a polished metallic mirror be used, no spot appears.

Since one of the glasses is convex, and the other plane, it is evident that their distance asunder increases from the point of contact, and that this is the cause of the colored rings. Now the curvature of the lens is known, and as it is determined that a given thickness of air corresponds to the length of waves of a given color, all that is to be done is to compute that thickness; i. e., the distance of the two surfaces apart. From the known curvature of the lens, the distance of the black ring is determined as 1-98000th of an inch, and from this the others are deduced. The red is found to be 0.0000.266 parts of an inch, the violet 0.0000.167 parts of an inch. Hence, by the law of undulations, the time of vibration is deduced. This law is, that the period of vibration to produce a given color is directly as the length of a wave of that color, and inversely as the velocity of light. Red light vibrates four hundred and fifty-eight millions of millions of times in a second, and violet seven hundred and twenty-seven millions of millions. White light is an aggregate of all the different waves.

Another proof that light is produced by waves is very easily obtained. If a sunbeam is allowed to enter a darkened room through an extremely small pin-hole, and received on a screen six feet distant, it will, instead of producing a white spot, produce a shadow surrounded by a series of colored rings, and dark intervals. When the screen is moved nearer, the central spot vanishes, and a succession of most brilliant colors appears. When one of the primary lights is employed, as red, the rings will be alternately red and black. The shadow of an object held in such a ray is bordered by colored rings. The cause of these rings is easily explained. The rays of light are bent around the edge of the object interposed, and pass onward to the screen in hyperbolic lines. The shadow is thus enlarged, and the rays of light become of different lengths, their waves in different states of vibration, and either produce the colored rings, or, interfering with each other, produce the dark intervals.

The colors of natural objects are produced by the interference or absorption of some of the colored rays. Green objects absorb all the rays but yellow and blue; red, all except those of that color. What becomes of the light thus absorbed? If the doctrine of emanations be received, this becomes a difficult question, for it has then to account for the extinction of matter; but if the theory of undulations be received, we have only to account for the extinction of motion. We can readily show how two opposing motions can produce an equilibrium, as two waves of water meeting produce rest; but to show how matter can thus become lost is a difficult problem. Some have entertained the idea that the light thus absorbed reappeared in the form of electricity and heat, and the intricate blending of the imponderables countenances the hypothesis; yet at least it is mere conjecture, plausible perhaps. Receiving the doctrine of vibration, we have, on the contrary, a real groundwork on which to rest.

The ear cannot perceive waves of sound unless they continue for a brief interval. So the eye cannot recognize a single wave of light, for it is gone too suddenly. A succession of waves must strike the retina to excite the sensation of sight, which is produced by these waves exciting vibrations in the retina, or the expanded extremity of the optic nerve, by which it is transmitted to the brain. This state of vibration does not suddenly stop when the cause is removed. Brightly luminous bodies are visible for the eighth of a second after they are removed.

Some animals can see in the darkest night, their eyes being sensitive to the feeblest luminosity; and perhaps some animals are sensitive to waves of light, to colors, and phenomena with which man is wholly unacquainted, as the instincts of some insects indicate.

It has been objected to the undulatory theory of light, that the waves do not follow the law of sound. We know that the waves of sound, when they enter a room through the smallest aperture, instantly pervade the whole apartment. On the contrary, those of light move in a solid ray across to the opposite wall. This objection is readily removed when we consider that the waves of sound are incomparably larger than those of light, and, although entering the smallest aperture, they expand and set the air in the room in vibration; whereas, of the united waves of light radiating from a luminous body, those which enter at the borders of the aperture diagonally, entirely destroy each other, while only those that pass directly in retain their illuminating power.

It is also objected,— and for a long time the objection was unanswerable, — that as the light from the sun and stars was white light, if composed of waves of different lengths, all must possess the same velocity; whereas their separation by the prism showed different velocities, either in space or the refracting medium.

M. Fraunhafer removed this objection entirely by producing a spectrum by making a sunbeam pass through a sieve of fine parallel wires drawn over the object glass of a telescope. In the spectrum so produced, the spaces occupied by the colors are dependent on the length of their waves; and he found that the distance between them was precisely as the differences of their lengths. He measured the length of the waves of different colors at seven different points; and Professor Powell, seizing the clew thus furnished, brought the subject under the immediate jurisdiction of mathematics, and produced a most triumphant proof of the wave theory.

Heat. — The solar spectrum does not consist of colors alone; besides the red, blue, and yellow rays, there are chemical rays, and rays of heat. The highest temperature is found between the red and violet rays; but it is not distributed continuously, but in spots analogous to the colors, and extends beyond the violet nearly one half on either side. Heat is always found in the spectrum formed by all luminous bodies. It enters the analyzing prism with the rays of light, and are affected in precisely the same manner; and if it be demonstrated that light is produced by waves, it is also demonstrated that heat is produced in the same manner. We see its separation from light as emitted from boiling water and steam, as we see one color separated from the others. Rays of heat, in their emission from heated bodies, follow the same law as light, passing off in straight lines from every part of the radiating surface. Rays of heat, from whatever source, are transmitted through solid or liquid substances, whether those substances are in motion or at rest, in a manner similar to light. But in most solid and liquid substances a remarkable difference occurs. While some of these, as thin plates of alum, transmit almost all the light, and intercept nearly all the rays of heat, others, as brown rock crystal, entirely prevent the passage of light, but freely admit the rays of heat. But even here a striking analogy is found to the transmission of light by colored media. When white light passes through a plate of red glass, all other waves, except the red, are arrested. As the rays of light from the sun have greater intensity than those from any other luminous source, so its rays of heat, when concentrated, produce a greater heat than can be produced by any other means whatever. They possess greater penetrating power; for while the heat of a common fire cannot pass through the thinnest plate of glass, it offers no resistance to the heat of the sun. Heat is also capable of being polarized, that is, having its vibrations reduced to one plane; and its circular and elliptical polarization have also been proved. From such data the inference is unavoidable that light and heat have a common origin; and if one be referred to undulations in an all-pervading medium, then must the other. Experiments show that the difference between them is in the length of waves, those of heat being considerably longer. That they can be separately produced does not argue a difference of cause; nor does their invisibility, which depends not on the waves of heat, but on the structure of the eye. The sense of seeing is narrowly confined. The waves of heat may be too rapid, or not sufficiently diffusive to affect the eye. Man is circumscribed in every thing. He is ignorant of the manner in which the antennæ of insects warn them of danger, or how the carrier pigeon finds its home though thousands of miles intervene, or how the vulture's eye rivals the telescope. Beings may exist to which an entirely new universe is revealed by different rays of light from those which affect the optic nerves of man; by different rays of heat from those to which he is sensible, and by sounds his ears are not capable of hearing.

The identity of effects, under similar circumstances, is most incontestable proof of the identity of cause of heat, light, and the chemical rays. All are capable of reflection from polished surfaces — of refraction — of polarization; they are destitute of appreciable weight; their velocity is incomprehensible; they can be concentrated or dispersed by convex lenses or concave mirrors; and they are equally capable of radiation.

The interference of waves of heat, by which both are destroyed, producing zero instead of redoubling the temperature, has been placed beyond a doubt, and completes the vindication of their common origin in the universal ether

CHAPTER VIII.

PHILOSOPHY OF THE IMPONDERABLE AGENTS IN THEIR RELATIONS TO SPIRIT, CONCLUDED.

General Consideration of the Solar Spectrum. — ELECTRICITY. Its Source. — Condition of an Electrified Body. —Velocity of Electricity.— Its Relations to the other Imponderable Agents. — MAGNETISM. Its Relations and Functions. — OD FORCE. NERVE AURA. Reichenbach's Investigations. How examined. — Crystals, Magnets, and Minerals, in their Odic Relations.— Correspondence of the Magnetism of the Earth and of Man. — Difference from Light, Heat, Electricity, Magnetism. — Proposed Classification.

THE rainbow is produced by decomposition of light by the descending rain-drops. Seizing this clew to the composition of light, philosophers proceeded to investigate, with the prism, a wedge of glass by which white light is decomposed and an artificial rainbow formed. A beam of light is allowed to enter a darkened room, and is projected on a white screen. If this beam is made to pass through a glass prism, instead of forming a white spot on the screen, a stripe of colors appears in its place, in the order they are seen in the rainbow — violet, indigo, blue, green, yellow, orange, red, lavender. Of these, but four can be considered as simple colors, the others resulting from various admixtures of the four primaries. Red, blue, yellow, lavender, are the four elemental colors, concentrated in stripes, but diffused in all parts of the spectrum. Besides the visible, colored bands of the spectrum, there are three others, the magnetic, calorific, and the as yet nameless one which exerts a magnetic or *odic* influence on animals and man. These are all invisible, and may be separated from each other, as heretofore shown. In

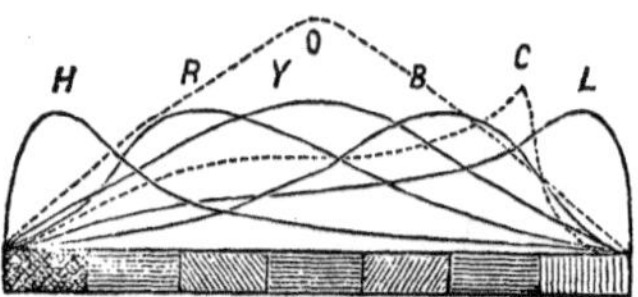

Fig. 1. Solar Spectrum.

the accompanying figure the bars represents the spectrum as seen projected by the prism on a screen. H represents the concentrated portion of the calorific rays,* which, however, are seen to extend over the whole spectrum. R represents the red, Y the yellow, B the blue, L the lavender rays discovered by Herschel, and satisfactorily proved to exist beyond the violet, being the most refrangible rays. C represents the chemical, and O the odylic, nerve-auric, or animo-magnetic. When we explain how all these rays are referred to a common source, as pulsations in a universal ether, their seeming complexity will be reduced to beautiful simplicity. Different length of waves, which produce these phenomena, explains all their collaterals, and shows how readily one can be evoked, or be converted into the other, as we can readily comprehend how a short wave may be converted into a long one, and *vice versa.* If the waves of heat are twice as long as those of light, then, if shortened one half, instead of operating on the nerves of sensation, producing the feeling of warmth, they would vibrate on the optic nerves as light. So, if light waves are longer than electric waves, if they strike and are absorbed by the earth, they may be shortened and circulate around that planet as currents of electricity governed by seemingly new laws, and producing a startlingly new class of phenomena.

* Instead of the inaccurate expression "rays" and "beams of light and heat," etc., as soon as the theory of the imponderables is explained, the term "waves," as more scientifically accurate, only will be used.

Or any of these may be converted into waves of odyle, producing the phenomena of life and animal motion, of which hereafter.

Faraday has proved the identity of electricity and magnetism by producing the spark, heating metallic wires, and chemical decompositions. But of its true nature philosophers have developed little by all their wranglings. Of the terms *positive* and *negative*, they may be employed in the arbitrary sense usually given them; but if understood as representing two fluids, or different states of the same fluid distinct and independent of the cosmical ether, they should be discarded as rather giving the air of philosophy to presumption, than belonging to true science.

An electrified body, like a luminous body, is one capable of exciting vibrations in ether — different vibrations, it is true, but of like nature. Here we have the key to all electrical phenomena. Induction, where one body produces a state of vibration in a contiguous body, is similar to that phenomenon of sound when one vibrating string sets another in vibration. The phenomenon of positive and negative is readily explained by the alternations of waves, so that the pulsations of one body exactly alternate with the other. Conductors are bodies producing vibrations readily; non-conductors, those which interfere with their passage. An insulated body electrically charged is in the state of the flame of a candle in an opaque box. The vibrations in either case cannot escape. Place the candle in a transparent, instead of an opaque box or a conductor in the place of the insulator, the vibrations get vent and escape.

The velocity of the electric vibrations is much

greater than those of light When a wheel, made to revolve with sufficient velocity to render its spokes invisible, is illuminated by a flash of lightning, it appears to be at rest, every spoke being visible; for however fast it may rotate, the light has already come and gone before it can revolve through a sensible interval.

Before we approach magnetism we must have a greater velocity of transmission than even the almost incomprehensible one of light, or electricity. Even that is too slow, for unless the vibration of magnetism is transmitted with still greater rapidity, an element of disorganization, as has been observed, is introduced into our cosmical system. If we grant electricity to be vibrations in the same ether as light, and that magnetism is identical in its origin, we have already before us the whole subject of gravitation. The attraction of particles in solution or affinity, the attraction of cohesion, attraction and repulsion, and gravity of worlds, are resultants of one common cause. If one world affects another, if the sun chains the planets in their orbits, there must be a bond of union between worlds. Something cannot originate from nothing. To call attraction a property of matter does not explain its *modus operandi.* The space through which suns and worlds float in an undeviating pathway is filled with an ether, and worlds and suns are great pulsating centres from which radiate the vibrations of attraction.

According to the Newtonian theory of gravitation, worlds, unless impelled by a constantly applied tangential force, would fall into their central suns. This view introduces an element of disorder into the system which is far from rational. On the contrary, if we adopt the wave theory, we can readily explain how worlds arrange themselves around their central orbs by the inter-harmonious relation between their pulsations; so that it is impossible for a world or rotating body to move from its orbit, as its repulsion and attraction only harmonize in that orbit; and if it was removed, it would speedily regain its natural position.

Thus have we glanced over the so-styled imponderables, and pointed out their interrelation. In case of either one of these elements, the supposition of a fluid cannot be entertained for a moment. They are all developed by friction, condensation, consolidation, vaporization, chemical action, crystallization, &c. Like music composed of notes or waves of different lengths, they are all resultants of waves of different lengths; thus from one cause widely diversified phenomena are produced. Their common source is beautifully illustrated by their reduction to mathematical accuracy and precision. All follow one law of decrease as the square of the distance. This is a necessary consequence of propagation by waves, as will readily be seen by inspection of the following diagram, where A represents the luminous point. Now, it is evident that the same amount of vibration passes between the lines A E and A D throughout their whole extent, and it is also evident from this that as the same vibration at C B spread over one surface is four times as great as at E D when it is spread over four surfaces. This property first discovered in light is possessed by all the imponderables. This is alone sufficient to reduce their effects to a common cause.

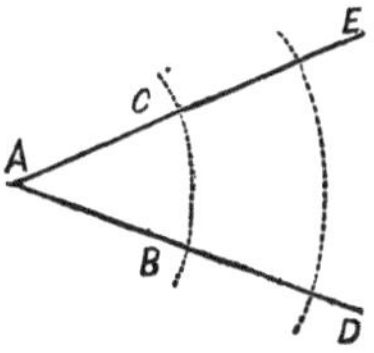

E
C
A
B
D

Fig. 2.

I have glanced at the imponderables in this cursory

manner, in order to approach more understandingly the most intricate of them all — that which is related to life, and is the cause of vital phenomena.

Baron Reichenbach, though far from being its discoverer, was the first to treat it scientifically and give it a name. Mesmer entertained the theory of a universal fluid, by which the phenomena styled mesmerism were produced; but he, as well as all others, did not venture to compare this medium with the space-ether by which all the imponderables are produced. From Reichenbach's work* I shall take whatever is relevant here, and once for all speak his immortal praise as the Galileo and Newton of this department.

Reichenbach, in his beautiful and accurate experiments, has proved that from the poles of the magnet a lambent flame constantly arises, plainly perceptible to the sensitive eye, and felt by the sensitive nerve. When the solid bar magnet is presented to a sensitive, a solid flame is seen to arise; but if the magnet is composed of layers, each layer gives out its own flame at a slight angle, but all mingle in a beautiful brush of light. This appearance is shown in the figures. Figs. 3, 4, are a five-layered magnet, presenting the tufted flame. Light is also emitted from the edges and side of the bars, but more feebly, as seen in Fig. 5. These remarkable facts were first discovered by Baron von Reichenbach, of Vienna, who, amid scoffs and sneers, continued experimental research in this direction, unti

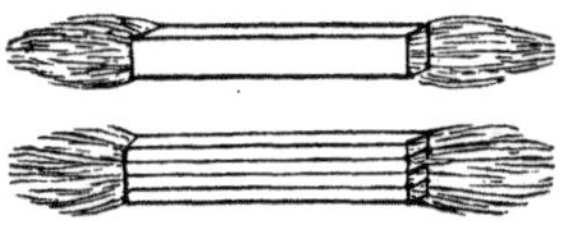

Figs. 3, 4.
Solid Bar, and Layered Magnet.

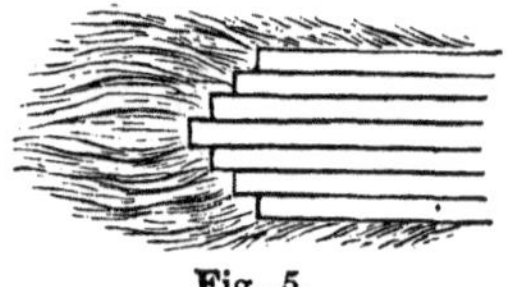

Fig. 5.

* Dynamics.

little remains in the way of his theory. Before, however, we enter into the discussion of this agent, we will state some of the facts by which he arrived at such important and unexpected results. His investigations were carried on by the aid of over sixty sensitives, some of whom were of the first ranks of society, and above deception. Even if so inclined, such was the care with which his experiments were instituted, that it would have been impossible. The subject was placed in a darkened room, and after becoming accustomed to the place, the magnet was presented. Madame Reichel saw the magnetic light not only in darkness, but in the dim twilight which was required to perceive all objects. In twilight the flame was smaller — that is, the borders of the flame were overcome by superior light, and she saw less of it; and the darker the room, the larger the flame, and she saw more of it. It is now well known that clairvoyants perceive this flame, and feel its influence; and a multitude of facts can be presented from experience of somnambulists, natural and artificial, to prove that such emanations exist.

From the foregoing we predicate that as all bodies are more or less magnetic, or electric, all must have these perceptible emanations. Crystals have a peculiarly strong influence, and act exactly according to their polarity. A large crystal to the clairvoyant or sensitive, presents a very beautiful appearance, which I

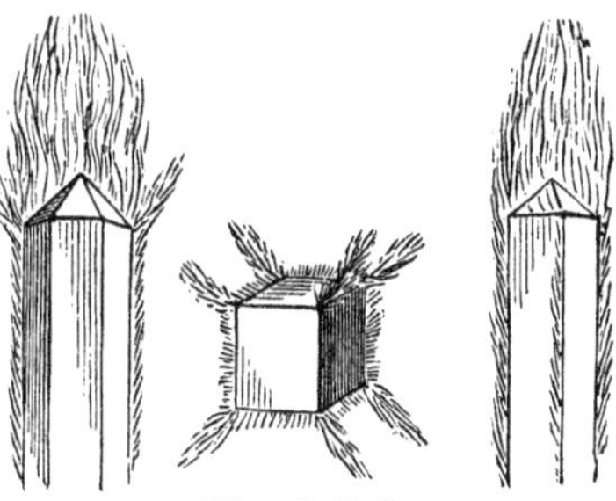

Figs. 6, 7, 8.

have vainly attempted to portray in the accompanying Figs. 6, 7, 8. No painter can do justice to the splendid play of the colored flames which dance around the primary and secondary poles. Countless experiments have proved the cause of the flame and influence to be universal. Whatever it may be we can only learn by observation of its effects. It is closely allied to magnetism, to heat, light, and electricity; yet in many respects it departs widely from these. When an iodized plate is placed in a dark box on the pole of a magnet, and allowed to remain a considerable time, the plate, when treated with mercury, exhibits the full influence of light. It is also concentrated by a lens in precisely the same manner as light. These facts show a close interrelation. It is moreover always present in burning bodies, flows in copious streams from the sun, moon, and stars, each of which has an influence peculiar to itself. It does not emit heat.

The flames from the positive and negative pole will not unite, wherein they display an affinity to magnetism; but its distinctness is shown in the fact that all objects rubbed with a magnet acquire this property, and act in precisely the same way as a true magnet except they do not acquire the property of attraction in the least degree. If a steel bar and a bar of lead are rubbed with a magnet, the former alone acquires magnetism, but both act with equal power on the nerves of sensation. The magnet imparts this property to all substances, even to living bodies.

This force, as found in the magnet, crystal, and living being, is identical in its properties, which may be briefly stated. It is conductible through all bodies. It can be accumulated or transferred; it shortly disappears or is dispersed, and attracts the sensitive as magnetism attracts iron.

There is a hidden relation between it and magnetism, as it will be observed that the direction and form of the flame exactly corresponds to the form and direction of the lines of iron filings gathered by a magnet; and it also circulates along the magnetic axis, and escapes from the poles of crystals, thus revealing their internal structure.

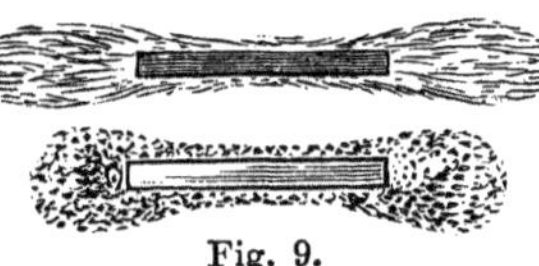
Fig. 9.

It is a force felt not only by contact, but at the distance of the stars. It is a universal force. It is related by positive and negative to electricity. All substances array themselves in an electro-chemical series — the electro-positives on one side, the electro-negatives on the other; but it differs in its universal conductility and its transparency, as in the diffusion of cold from the positive and warmth from the negative poles.

The fluctuations of the magnetism of the earth and of man remarkably coincide. The sensitive are restless, unless they repose in a position corresponding to the magnetic meridian. Reichenbach speaks thus: "When I tried Miss Nowating (a sick sensitive) with the magnetic needle, I found her almost exactly in the magnetic meridian, her head (positive) towards the north. She had herself sought and wished for that direction; and it had been necessary to take down a stove to satisfy her desire. I asked her to lie with her head to the south, by way of experiment, to ascertain the result. It required some pains to induce her to do it; for I was obliged to repeat my wish three or four days successively, and to make her appreciate the weight I laid on the change At length I found her one day in this reversed position. A little time elapsed before she began to complain. She was uncomfortable; she turned over restlessly; her face became flushed; her pulse rose, became feebler, and the flow of blood to the head increased; headache and discomfort of the stomach ensued. Her head was quickly turned, but stopped at a quarter of a circle. But her discomfort did not cease until she reached her former position, the one she sought by instinct, as the needle turns to the pole, when all these disagreeable symptoms vanished, and her nerves were restored to their former tone."

The generalization from such facts is thus expressed:* "*The terrestrial magnetism exercises in sensitive persons — healthy or sick — a peculiar, exciting action, strong enough to interfere with their rest; in the healthy, to modify their sleep; in the sick, to disturb the circulation of the blood, the functions of the nerves, and the equilibrium of the vital force.*" Here is seen the interrelation between animal magnetism and that of gravitation. Now we will show how closely attraction, as manifested in worlds, is identified with the force resident in the animal body. "Professor Endlicher advised the physician to pass the magnet over himself, and then to react on the patient, (instead of using it directly.) To his surprise, he now, as had never happened before, could attract the hand of the patient with his hand, cause it to attach itself, and follow every where just as the

* Dynamics of Magnet, p. 100.

magnetized glass of water had done. He retained this power for almost a quarter of an hour; by that time it had by degrees dropped. The same unknown something, which had been left in the iron rod by the magnet, and had likewise passed into the glass of water, must, therefore, have been conveyed into the whole person of the physician. "When I passed a magnet down twice from head to foot, over the patient, she lost consciousness, and fell into convulsions, mostly with rigid spasms. When I did the same with my large rock crystal, the same result followed. But I could produce the same effect, when, instead of either of these, I used merely my unaided hand."

Closely related, however, as animal magnetism may appear, sufficient distinctions exist to render its admission among the imponderables as a distinct species necessary. From heat it is distinguished by not influencing the thermometer in the least; even Nobili's thermoscope remaining motionless in the focus of a concentrating lens. At the same time it affects the sensitive precisely as heat, producing a burning sensation on the one hand, and of cold on the other. Heat and animal magnetism are often directly opposed in their effects. The right hand produces the sensation of cold on the sensitive, while it always indicates heat to the thermoscope. The rays of the sun, always warm to the thermometer, are cold to the sensitive. Moonlight, always cold, produces a sensation of warmth. If combustion and chemical action always liberate heat, they always produce the sensation of cold on the sensitive.

Bodies, which are non-conductors of heat seem perfectly permeable to magnetism,* as glass, rods of wood, &c.

Heat always changes the volume and density of bodies it pervades; but magnetism does not perceptibly change the volume of bodies it permeates.

From electricity it is distinguished by penetrating the mass of matter; whereas that element only manifests itself on the surface. It can be accumulated on any object, whether insulated or otherwise. It has not that extreme tension or tendency to diffusion which renders the other impossible to be retained long, even with the best insulators. When thus accumulated, it cannot be annulled immediately, but hours and days are required for its discharge. On the other hand, electricity discharges itself in an inappreciable part of a moment. All continuous bodies are almost equally good conductors of Od; but it is conducted slowly, whereas few are free conductors of electricity, and it passes almost instantaneously.

Electricity often excites Od. The electrical machine is electrically excited for some time before the odic flame becomes visible. But on the other hand, it continues indefinitely longer. Sometimes the reverse happens; so that it is evident that although electricity excites Od, the latter afterwards pursues its own course.

Odic or magnetic flames never unite and neutralize each other.

Such facts conclusively show that there is a true distinction between these two elements.

The distinctions between what is usually known as animal magnetism and the attraction of the magnet

* Animal magnetism, Od, or nervous fluid, always understood.

are equally clear. Animal magnetism manifests itself in a number of cases where magnetism does not exist. It is "present in many chemical processes, in vitality, in crystals, in friction, in the spectra of the sun, moon, and candle light, in polarized light, and the amorphous motional world generally." Although animal magnetism occurs without any magnetism whatever, the latter never occurs without the former. While the existence of the latter in moonlight and sunlight is doubtful, the former exists in vast quantities. A cloud obscuring the sun almost destroys its effect on the sensitive, whereas magnetism is not weakened by the interposition of any substance whatever. Animal magnetism can be transferred to all bodies, whereas magnetism is very partial, and few substances are capable of retaining it. The former is conductible by all substances, and capable of reflection and refraction; the latter is not.* Magnetism, like electricity, is distributed on the surface; the force under discussion pervades the mass; and while it powerfully attracts the sensitive, it is incapable of supporting the smallest iron filing, and bodies charged with it do not take a north and south direction, although something like this is seen in the highly sensitive, who prefer reposing with their heads to the north.

Distinguished thus from, and related to, electricity, heat, light, and magnetism, the universal fluid of Mesmer and his school assumes a new form, and takes its place beside the imponderables. So far we have employed the generally received term "animal magnetism;" but we have now identified the cause of all the diverse phenomena referred to the imponderable agents.

* L'Institutes, 1846, p. 647.

We find they are a unity. One cause produces all the effects we observe, and scientific accuracy requires a classification. The old method of designating each separately is rude and bungling; a generic or family name must be used common to all, and a specific name given to each. A generic name is found in ETHER—the common medium of vibration—and employing it as a termination, and prefixing the descriptive name, we have a classification reading thus: —

Chromether (Greek *chroma*, color) — Light.

Calorether (Latin *calor*, heat) — Heat.

Electronether (Greek *electron*) — Electricity.

Magnetether — Magnetism, Attraction, Gravitation, Affinity, &c.

Zoether (Greek *zoe*, life) — Neur-aura, Animal Magnetism, Nervous Fluid, &c.

Perhaps it is premature to hazard such a classification; yet it presents the subject in much better form than the usual method of independent designations, which show no interrelations between these blended effects. Perhaps this classification may not be adapted; but if not, it will lead to more successful attempts. At least, with a consistent theory and classification before him, the student can better understand and apply facts, and lead off, by their aid, into new and brighter fields of investigation.

THE HISTORY AND LAWS OF CREATION.

CHAPTER X.

THE IMPONDERABLE AGENTS AS MANIFESTED IN LIVING BEINGS.

Suns. — Pulsating Hearts. — Light. — Heat. — Electricity. — Magnetism. — Zoether in the Relations to Life and Inorganic Nature. — Electrical Fishes. — General Considerations.

THE imponderable agents produce the most sublime phenomena of the external world. Light flows from the countless suns of the universe in a vast deluge, and its waves fly onward with inconceivable velocity, only expiring on the coast line of space. The luminous vibrations from all worlds commingle as they rush onward. Every sun is a great pulsating heart, from which these undulations flow as an irresistible flood. Here we discover a strong proof of undulation and argument against emanation; for it is difficult to account for the origin of so much matter as would be thrown off, as light and heat, by suns, whereas motion is readily accounted for.

How active is light in the organic world! The plant dies if deprived of sunlight, and expands its broad leaves beneath its genial rays. The animal perishes in darkness; its multifarious functions go on only in light. What joy, what happiness in light, flooding the world like a downy envelope, in which animated nature buds and blooms!

Heat, whether glowing in the grate, or beaming in

the light of the sun, is ever welcomed by the world of life. The earth basks in the solar warmth, as it rolls its teeming sides towards that luminary, a new life awakened in its bosom. The calorific breath sets in motion electric and magnetic currents in the earth. The needle swerves from its true place, and indicates the disturbances in the internal forces of our planet, and, perhaps, at night the fettered agents seek an equilibrium, and we are astonished by the northern heavens becoming a waving sea of fire.

Likewise does it awake the plant and the animal to life. The dead live, and the slothful become filled with activity. The sap ascends the trunk, the blood courses along arteries and veins, and all the allied agents are awakened.

Electricity is seen in the lurid stream of lightning coming and going in the millionth part of a second, and hoarsely speaking in the jarring thunder. More silently, but far more herculean its labors, than in the fierce crash of the tempest. Deep in the earth it deposits metallic veins, crystallizing minerals in forms of beauty, decomposing and recomposing in endless permutation. In the animate world it kindles the spark of vitality.

Magnetism holds the moon to the earth, the earth to the sun. It chains secondaries to primaries, primaries to central orbs, compels them to rotate on their axes, and revolve round common centres, and having thus unitized a system, it sends it on around a grander centre, around which millions of systems as complicated are borne like flocks of down on the eddying winds of autumn.

Neur-aura, Od force, or Zoether, — the term we have adopted to express this class of phenomena, — the life emanation, produces in the animal similar effects as magnetism does in worlds. It is emphatically the life-force. By it, thought, that mighty engine which curbs the elements of nature and binds them in abjectest slavery, is manifested, and through and by it man becomes an immortal being, a kindred of the gods, rising from sphere to sphere in eternal progress, while suns and solar systems crumble and melt away like the mushrooms of an hour.

Such are the forces of which we treat — a subject sublime in its aspects, almost impossible to approach. We can but fail in the ultimate extension of our research, for the universe is infinite and our observations imperfect. But we may enter a little way into the vestibule of these dark mysteries.

In the previous volume we defined a living being as an individualization of the living principle of nature. We said that the same forces which exist in the universal world, if individualized, become a living being. Thus it becomes a *centrestance*, as well as a circumstance, and from it forces are propagated as from a creative cause.

Here opens a beautiful field of investigation, in which beauty and sublimity blend in the delicacy of microscopic organizations, and clairvoyant sight.

Light is emitted by most organized substances in decay, and by many living beings. The human subject is sometimes self-luminous, and this is far from an uncommon occurrence in the inferior orders of living beings. The growing mushroom sheds a clear phosphoric light; the firefly attracts its mate by its brilliant lamp; and often is the illimitable ocean converted into a rolling sea of flame by the light of countless swarms of its microscopic denizens.

Heat is characteristic of the organic world. It is inseparable from change, whether mechanical or chemical. Even the plant emits a sensible heat, and cold-blooded creatures, which always remain at the temperature of the medium in which they exist, produce a sensible degree of caloric. In warm-blooded animals changes occur with great rapidity, and their bodies are maintained at about one hundred degrees, whatever be the temperature of the medium in which they are placed. They originate heat like the sun; that is, the changes which take place in them produce vibrations.

What are these changes? They are too numerous to catalogue, but a few of the most prominent may be profitably introduced. The great source of animal heat is the conversion of carbon, taken as food, into carbonic acid by the lungs, when, after yielding its heat to the blood, it is expelled. This is only a different form of combustion, but is essentially the same as that of wood or coal in the grate. Oxygen combines with its equivalent of carbon in both instances, and the latter is converted from a solid to a gas, thus producing vibrations in the universally diffused ether. There is the conversion of solids to fluids, and both to gases, and the reconversion of gases into fluids and solids; all of which changes, constantly occurring, are accompanied by the manifestations of heat and electricity.

The muscles have a difference between their internal and external portions of what is called positive and negative, and also between their extremities. In short, by the peculiar arrangement of organs, the living body becomes a galvanic battery. The arteries and veins permeate the muscles in a perfect mesh of capillaries. These are filled with blood, an alkaline fluid. On the other side of their delicate walls is the acid fluid of the muscles. Thus a constant battery is formed; but as no special apparatus is formed for the reception of the electric fluid, it passes away as fast as formed. Cases are recorded where the human patient became so electrical as to be enabled to charge Leyden jars, and produce sparks of considerable intensity. But no special agent exists in man for its manifestation, as in the electrical fishes. Of the seven genera known, belonging to widely separated families, the gymnotus of the South American rivers is far the most powerful. Its shock is sufficient to kill small animals and fishes, and horses driven through the rivers, and receiving a succession of shocks, are said to be frequently killed. It is very painful to man.

Some psychologists have considered the pure electricity thus manifested as the real neur-aura, or nerve fluid; but essential differences exist between the two. Zoether — neur-aura — is a distinct manifestation, perhaps secondary, perhaps primary to all the others. We, however, only treat of them sufficiently to the understanding of the views we present, the greater portion of the volume being devoted to the discussion of the zoethic manifestations.

Thus we find that living beings manifest a modified

form of the *inorganic imponderable agents.* Zoether, as its name imports, belongs wholly to the region of organic forms. It is the atmosphere of the spirit, and we shall find, as we proceed in our investigation, that it forms the basis of spirit existence.

CHAPTER XI.

SPIRITUAL ELEMENTS.

Infinite Progress of the Elements.—Soil of Granite Mountains, Potash as applied to; Phosphorus applied to.—Infinite Variety of Matter.—Line of Demarcation between the Imponderable Agents and Spiritual Elements.—Philosophy of Organic Attractions.—Spiritual Elements, their Character and Functions.

ETERNAL progress is written in the constitution of matter. Every new form it enters, every elimination from an old, produces a more refined state. True, this may not be perceptible to chemical tests; but there are other methods by which it can be proved. The spongioles of the rootlets of plants make a more delicate analysis than the most experienced chemist. By reference to the first Volume, it will be seen that ashes are the best fertilizer that can be applied to the soil of decomposed granite, such as forms on the flanks of granite mountains. This soil is saturated with potash, yet is sterile, and only becomes fertile by application of potash derived from ashes. The chemist pronounces potash derived from ashes and from granite rock identical; the plant says that a wide difference exists between them. The moss and the lichen, the lower forms of vegetables, can adhere to the side of the rock and absorb its particles as fast as they decompose; not so, however, with the higher forms of plants. They cannot readily assimilate the crude potash, but require, before it becomes food for them, that it should pass through other vegetable forms. This fact is a clear annunciation of the progress of the elements. Another is drawn from the same source. Phosphorus, as existing in bones, is one of the most stimulating fertilizers known. It also is found in one variety of limestone rock laid down in the early ages of the world in vast quantities; but repeated trials have proved that although the phosphorus from both these sources is identical by the tests of chemistry, while the first is more valuable than gold, doubling and trebling the harvest, the latter is utterly worthless, producing no perceptible effect on vegetation. Is it not at least a plausible deduction that this element, as derived from these different sources, is in itself different? One has remained free, and passed through living organisms myriads of times; the other was laid down in the rocks in early ages, before it had departed from the crude original of the abysmal ocean of the beginning. Hence one is refined; the other gross. So speaks the living plant when the analysis is made by its spongioles.

If, by passing through living organisms, matter is refined, between the two extremes here stated, all degrees of refinement must exist; for each absorption, each elimination, must change the particles acted on; hence all varieties of matter must exist. And as the same process continuing for coming ages must produce precisely the same effects, we must no longer regard this planet as composed of sixty-four primary elements, or of any given number, but as formed of an infinite variety, in all degrees of refinement. There are all grades of oxygen, of carbon, &c. Living beings are formed of these. Each organized being draws to itself particles which affinitize with itself, or for which it has attraction, and repels all others. We see that such must necessarily be the case. If sugar, oils, starch, and allied alimentary materials, were absorbed or assimilated by animals in an equal degree, whatever be their sources, why is it that so large quantities are devoured while so small an amount is assimilated? The full-grown animal or man devours ten times the amount of food that is actually used in repairing the waste of the system; the remainder is rejected, because unfitted to enter the system, or because there is no attraction between the refused particles and the organism. Yet to the rough tests of chemistry the food may be identical.

Like attracts like. Particles once incorporated into the organism, attract others of like character. An individual inclined to corpulency may eat the same food and in the same quantity, as one in the other extreme of spareness, and while the latter grows leaner, the former will increase in *embonpoint.* One system rejects, the other retains, the fatty particles of food. We ascend a step higher. The higher always governs the lower. Attractions and repulsions are ruled by more refined elements.

Here let us draw a sharp line of distinction between those agents we have previously treated as imponderable, the laws of which we have attempted to explain, and the spiritual elements of which we at present treat. They are distinct, and are allied only by their refinement. Their properties are distinct, and their offices widely separated. It would be impossible to form an organism from these agents. Subtile, repellent, infinitely diffusible, they are instruments for us, not elements to build up our organizations. The infinite repulsion of the imponderable alone renders it useless as an element of organization.

The spiritual elements, such as the earth emanates, which go to form the spiritual spheres and enter into the organization of spirits, are realities. They possess all the properties of earthy matter, with new ones which they acquire by their refinement. Carbon is represented by a spiritual carbon, oxygen by a spiritual oxygen, &c., through the long catalogue. Hence we can be organic beings as much as while on earth, and our organs can perform their functions, and be supported by elements appropriate to those functions.

Another explanation concerning the unindividualized beings whose spiritual essence ascends into the vast ether, and gravitates like an evaporating cloud to its appropriate position, is here afforded. True, they are not individualized; they do not retain their identity; but they again enter into somewhat similar forms. If of sufficient refinement, the atoms pass at once to the spirit sphere; if not, they reunite with gross matter, and enter again the cycle of living beings, to be again and again eliminated, perhaps to travel up to the human form divine, and becoming embodied, stand forth as eternal as the everlasting planets — nay more, when these shall fade like the baseless fabric of a vision, rise above the wreck of worlds, rejoicing in increasing wisdom.

One law of attraction and resulting repulsion exists both in the earthy and spiritual spheres. The poison wolf'sbane twining its roots around and among those of the fruitful corn, extracts from the same dew, the same rain, the same soil, the most deadly poison, while the corn elaborates the life-giving grain. Particles seek like particles. They are repelled from dissimilar ones, and thus the intricate and mysterious web of nature is woven.

So in the spiritual world. The same law rules supreme. The force which builds up the wolf'sbane and the corn, side by side, builds up from the ascending atoms the orange and the vine which decorates the landscape of the spirit-spheres.

We utterly discard the usual classification of spiritual elements which places the imponderable agents with them, or makes them material elements. Separated from the realm of the ponderables by the infinite repulsion of the atoms of the ether in and by which they are manifested as effects, they cannot, nor must not be employed in philosophical discussion as material agents. By so doing, confusion is introduced into the spiritual domain, and the idea of introducing broad generalization becomes utterly hopeless. We have endeavored to show what a broad generalization covers the field of these agents, and we shall find equally grand ones pervade the truly spiritual domain.

CHAPTER XII.

ANIMAL MAGNETISM, (ZOETHISM,) ITS PHILOSOPHY, LAWS, APPLICATION AND RELATION TO SPIRITUALISM.

Sympathy. — Illustrations of. — Animals can influence Animals. — Man can influence Animals. — Animals can influence Man. — Man can influence Man. — A common Cause for these Phenomena. — Exploded Objections. — Referable to Zoether, (*nerve aura.*) — Animal Magnetism. — Proofs. — *Impressibility of the Brain.* — Psychometry. — Its Laws. — Doctrine of Spheres. — Zoethism. — Body and Mind mould each other. — Psychometry. — Its Relation to Animal Magnetism. — Estimate of the Number of Susceptible Persons. — How known. — Choice of Tests. — Application to the Sciences.

A MYSTERIOUS sympathy exists between all living beings. Attraction and repulsion are exerted as well by animals as man. The swarms of medusæ in the ocean congregate by the same law as shoals of fishes, herds of bisons and wild horses on the western prairie, or man in the complex relations of society.

Love is a fervent manifestation of the same principle. The north and south, the male and female, principles attract. So from lowest to highest; from the mineral atom to the living being; from the cellular *protophyte* * to the thinking man, attraction and repulsion rule with iron sway.

When strange herds of animals are mingled, how soon they separate if left to themselves! So in society, how kindred spirits unite in bonds of friendship, and the bad avoid coming in contact with the good!

We know of no instance better illustrating the sympathy existing between all individuals than the Si-

* See vol. i.

amese twins. They furnish an overdrawn example, it is true, but trace the law of sympathy with lines the more sharply defined. So intimately are they related, that the thought of a surgical separation, which was suggested to them, was met with as much horror as the idea of losing half of the body by the same means would be by the normal individual. The operation, if performed, would, from the closeness of the sympathy existing between them, undoubtedly prove fatal. Their hunger, thirst, sleeping and waking, are coincident, and their tastes and inclinations the same. They read the same book, and both play the same game, but never with each other, as they say that that would be like the right hand playing with the left. When one is sick the other has precisely the same symptoms. So simultaneous are their movements, that it is impossible to decide from which the impulse originates.

They are in a similar relation to each other as the fœtus and mother, between which a sympathy is established, which blends both into one entity, and transmits the slightest shade of thought from the mother to the offspring, often stamping the plastic being with the impress of her sensations as by the inexorable decree of fate.

The same sympathy is often shown by persons twin-born. Instances are recorded in which, although at a considerable distance from each other, the same malady appeared in both at the same time, and ran precisely the same course.

A young lady was suddenly seized with an unaccountable horror, followed by convulsions, which the attending physicians, unable to account for, said exactly resembled the struggles and sufferings of a person drowning. Soon after news came that her twin-brother had at that identical moment fallen overboard and been drowned.*

A strong sympathy also exists between parents and children, husband and wife, and intimate friends, so that when one is in trouble or misfortune, the other becomes conscious of it. This is too well known to require an extended statement of facts. How often do husband and wife think the same thought at the same time, or answer the same question in the same manner. This occurs far too often to be referred to coincidence. A deep principle underlies it.

Very often persons who are unimpressible when awake are impressible in sleep, and then are conscious of this sympathy with others. In illustration of this proposition, one instance will be introduced as a sam-le of its class.

A Scottish gentleman dreamed that on entering his office in the morning, he saw a person, formerly in his service as clerk, seated on a certain stool. On asking him the motives of his visit, he is told the circumstance which brought the stranger to that part of the country, and that he could not forbear visiting his old master, and passing a short time in his former occupation. In the morning, on entering his office, he finds his dream prove true to the letter. Here the sympathy between them was great, and the ardent thoughts of the clerk impress themselves on the master. Perhaps it may be considered more probable that this dream was the result of a spirit impression — that the spirit guardian of the clerk went to the couch of his master and whispered this dream, as a preparation for the meeting; but

* Night Side of Nature, where a mass of such facts may be found.

eager as we are to prove spiritual intercourse, we must throw aside all this class of facts, which depend on the same law, it is true, as spirit communion, but have a mundane origin.

How often do we hear, when entering a company, however unexpectedly, "Ah, we were just speaking of you!" and the same is embodied in the old proverb, "The devil is near when you are talking of him." Our emanation, or sphere, reaches our destination before us.

If we trace the relations of this sympathy, we shall find that—

1. Animals can influence animals.
2. Man can influence animals.
3. Animals can influence man.
4. Man can influence man.

1. *The Influence animals exert over each other.*—"Professor Silliman* mentions that, in June, 1823, he crossed the Hudson at Catskill, in company with a friend, and was proceeding in a carriage by the road along the river, which is there very narrow, with the water on one side, and a steep bank on the other, covered by bushes. His attention at that moment was arrested by the number of small birds, of different species, flying across the road and then back again, and turning and wheeling in manifold gyrations, and with much chirping, yet making no progress from the place over which they fluttered. His own and his friend's curiosity was much excited, but was soon satisfied by observing a black snake, of considerable size, partly coiled and partly erect from the ground, with the appearance of great animation, his eyes bril-

* Quoted in Fascination, p. 14

liant, and his tongue rapidly and incessantly brandishing. This reptile they perceived to be the cause and centre of the wild motions of the birds. The excitement ceased, however, as soon as the snake, alarmed by the approach of the carriage, retreated. The birds did not, however, escape, but rested on the bushes, probably to await the reappearance of their enemy." It would seem that they were magnetized, and did not immediately recover, or they would have flown away, instead of remaining near the scene of their fright.

A story is told by a gentleman in Pennsylvania, who, returning from a ride, espied a blackbird describing circles, gradually growing smaller, around a large black snake, all the time uttering cries of distress. As the bird almost reached the open jaws prepared to receive it, the gentleman drove away the snake, when the bird flew away, uttering a song of joy.

Another anecdote is related of a ground-squirrel. He was observed running back and forth along the trunk of a large tree, his returns being each time shorter. The observer at length saw that the cause of the squirrel's peculiar movements was the fascinating influence of a large rattlesnake, the head of which was thrust out through a hole in the trunk of the tree, which was hollow. The squirrel at length gave over running, and laid himself down near the snake, which opened its jaws and took in the head of the passive squirrel. A blow across the neck of the snake caused it to draw in its head, and the squirrel, thus released, frisked away with the utmost precipitation. Dr. Good observes this singular power of the rattlesnake, and it is probably possessed by all the larger kinds of snakes. The fascinating serpent appears to exert an irresistible influence over its victim, which cannot for a moment avert its eyes from the object of terror.

Borrows informs us that, while travelling in the interior of Africa, he saw a large serpent in the very act of fascinating a bird — a species of shrike. The bird, when first observed, was at some distance from the snake; but it gradually approached, apparently irresistibly drawn towards the fiery eyes and open jaws, trembling convulsively, and uttering piteous cries of distress. He shot the snake, but the bird did not fly away. On approaching he found the bird dead, although it had not approached within three feet and a half of the snake. This fact shows how intimate the relation of fascinator and the fascinated is even in animals. The narrator supposes the bird died of fright — a most inadequate explanation. We know that the magnetized enters into all the phases of feeling of his magnetizer, and whatever affects the latter equally affects the former. The sudden death of the serpent would by sympathy shock the bird in an equally great degree if the influence was perfect.

Another anecdote is told * of a mouse being placed in the cage of a female viper. It was at first greatly agitated, but in a short time drew gradually near the viper, which, with fixed eyes and distended jaws, remained motionless. It continued to approach, and at last ran into the viper's mouth and was devoured.

2. *Man can influence Animals.* — Bruce, the African traveller, speaks in the most positive manner of the power the blacks of Sennaar exert over the most poisonous serpents, against which they seem armed by

* Philosophical Transactions.

nature. They take the horned serpents in their hands at all times, put them in their bosoms, or throw them at each other like balls. The influence exerted on them is so great that they scarcely ever attempt any resistance, and when they are irritated to bite, no inconvenience arises even from the fangs of the most poisonous serpents. "I constantly observed," said he, "that, however lively the viper was before, upon being seized by any of these barbarians it seemed as if taken with sickness and feebleness, frequently *shut its eyes*, and never turned its mouth towards the arm of the person that held it." They are often so debilitated by this fascination as to perish as certainly, though not as speedily, as though struck by lightning.

A gentleman residing at Oxford had in his possession a young Syrian bear, from Mount Lebanus, about a year old. This bear was generally good-humored, playful, and tractable. One morning the bear, from the attention of some visitors, became savage and irritable; and the owner, in despair, tied him up in his usual abode, and went away to attend to his guests. In a few minutes he was hastily recalled to see his bear. He found him rolling about on his haunches, faintly moving his paws, and gradually sinking into a state of quiescence and repose. Above him stood a gentleman well known in the mesmeric world, making the usual passes with his hands. The poor bear, though evidently unwilling to yield to this new influence, gradually sunk to the ground, closed his eyes, became motionless, and insensible to all means used to rouse him. He remained in this state for some minutes, when he awoke, as it were from a deep sleep, shook himself, and tottered about the court, as though laboring under the effects of a strong narcotic. He exhibited evident signs of drowsiness for some hours afterwards. This interesting scene took place in the presence of many distinguished members of the British Association, when last held in the University at Oxford.

This power is used by man to disarm the fury of the most savage animals. Robbers have learned and exercise this art on watch-dogs, the most furious of which they reduce to silence. The Laplanders exercise the same power over their dogs.*

It has become a matter of history that man can, by his will, control the horse and other domestic animals. Alexander taming his Bucephalus is paralleled in modern times by Sullivan and Rarey. Sullivan, in an hour's time, could so magnetize the most furious horse, as to make him follow and obey like the best trained dog. Rarey at the present tames, in the same manner, not only horses, but that untamed steed of the desert — the wildest, fiercest, and most unmanageable of the equine race — the zebra. The lion and tiger are fawning as kittens beneath the gentle yet inexorable sway of this influence. From the human eye a power goes forth, which, when rightly employed, controls the most savage beasts.

3. *Animals can influence Man.* — There are well-attested instances of animals exerting a mesmeric influence on man. A gentleman, while walking in his garden, was attracted by a snake he accidentally saw in the bushes. He watched it closely, and soon found himself unable to draw away his eyes. The snake

* Lindencrantz.

appeared to increase immensely in size, and assume in rapid succession a mixture of the most brilliant colors. He grew dizzy, and would have fallen in the direction of the snake, had not his wife at that instant come to the rescue.

Two men in Maryland were walking together, when one found that his companion had stopped by the road side. On turning, he perceived that his eyes were fixed on a rattlesnake, which had its head elevated and eyes glaring on him. He was leaning towards the snake, and crying, "He will bite me! he will bite me!" "Sure enough, he will," said his friend, "if you do not run off. What are you staying here for?" Finding he could not draw him away, he struck down the snake with the limb of a tree. The man thus saved was very sick for some hours afterwards.

4. *Man can influence Man.* — The influence man exerts over man was well known to the ancients. The physician's prescription for King David in his declining years was based on this principle. And the rites of savage nations, the gestures of the magicians and medicine-men over their patients, are founded on mesmerism, and are remarkably successful.

The facts of mesmerism are almost universally admitted by thinking men, and to introduce a list of established incidents would be superfluous. The reader is referred to the best works on that subject* for evidence.

The bond which unites the magnetizer and the

* Townsend's Facts in Animal Magnetism; Deleuze, Introduction; Gregory's Letters; Deleuze, Critical History; the various volumes of the London Querist; and the Journal of Man.

magnetized is illustrated at every successive process. The magnetizer exerts despotic sway over his subject. He compels him to think and to act as he pleases. If he tells him water is wine, he is implicitly believed, and intoxication follows. He tells him a stick is a snake — he flees from it; or that he is a king, or emperor, or czar, and the character is assumed; or he plunges him to the opposite extreme, and he crawls along, a degraded outcast. In these fantasies, if we may so call what depends on the will of another, there is something strikingly similar to the operation of narcotics, especially *hashish*, or Indian hemp, which the Hindoos used to produce the ecstasy in which they communicated with the gods, and learned the course of future events. In short, there is very little distinction to be made between the effects of the narcotic and of the magnetizer.

Whatever affects the magnetizer affects equally the subject. The slightest pain, the least desire, is participated. Whatever the former tastes, hears, or sees, the latter tastes, hears, or sees; and there is a partial reflex action by which the magnetizer is guided to the locality of, and sympathizes in the diseases of, his subject.

The following often quoted instance is taken from the Transactions of the French Academy.

" On the 10th of September, at ten o'clock at night, the commission met at the house of M. Itard, in order to continue its inquiries upon Carat, (their mesmeric subject;) the latter was in the library, where conversation had been carried on with him till half past seven, at which time M. Foissac, (the magnetizer,) who had arrived since Carat, and had waited in an ante-chamber separated from the library by two closed doors and an interval of twelve feet, began to magnetize him. Three minutes afterwards, Carat said, 'I think Foissac is there, for I feel myself oppressed and enfeebled.' At the expiration of eight minutes he was asleep. He was again questioned, and answered us," &c.

A friend says that there is a member of the family in which he resides who oppresses him whenever near. No enmity exists between them; but their organizations are entirely different.

[One evening, while engaged in conversation with Dr. B——, he suddenly paused, and said he could proceed no further, for some one was listening. This was highly improbable; but the next morning tracks were discovered at the gate, in the light fall of snow, as though some person had stopped for a considerable time.]

Whatever influence that person exerted must have passed through the park, yard, and wall of the house, to reach the impressible brain. Shall we call it thought? What is thought? How does it reproduce itself in the mind of another? These are questions which force the close reasoner to the adaptation of an ethereal medium of transference.

These curious phenomena have long been observed and speculated on. To extend the list is unnecessary, for almost every one can recall equally conclusive facts.

One thing is determined — they do not arise from imagination, for we see phenomena in animals that cannot be referred to imagination. It is possessed by animals as well as man. Animals influence man

man, animals; animals, each other; and man controls man.

To produce a result so uniform, we must assume the cause to be common to all. Hence we refer this entire class to ZOETHER, or what, perhaps, will be better understood, nerve-aura, in which all living beings can excite undulations.

In the world of mind, theories have ever gone before, as pathfinders, so to express it, long before sufficient facts are gathered for their support. So has it been in the present instance, in an eminent degree. The existence of a *nerve-aura* has been maintained and denied by eminent psychologists, but the affirmative have considered it as an emanation — a theory which soon leads to its own destruction.

We have already shown what is to be understood by *nerve-aura* and kindred expressions. The nervous system is capable of exciting or conducting these vibrations. It has been said that the nerves are non-conductors of electricity, and it has been supposed that this fact alone destroyed all theories hitherto entertained of the subtile influence persons exert on each other. But we cannot perceive how this fact is related to the subject. We well know that dynamic electricity, or that generated by a battery, must have a closed circuit, and that water, the principal ingredient in the composition of nerve, is a bad conductor of this species of electricity. But water is a good conductor of static electricity, or that generated by the electrical machine. So that it is false that nerves will not conduct electricity. Further, it is zoether, and not electronether, of which we treat, and our experiments must be with it, and not with any other form of ether. We must consider all the facts we have been and are considering as isolated phenomena, or unitized by this zoether.

From remotest antiquity the adage has descended, that young people were in danger of becoming unhealthy by living with the aged. The Hebrews acted accordingly in procuring a young damsel for their old king, that he might be invigorated by her strength. An anecdote is told of an aged female who compelled her servants to retire in the same bed with herself, that she might prolong her life thereby, and carried her horrid vampyrism to such an extent that, her maids all becoming sickly, after a time, she could induce none of them to work for her, and soon expired. This explains why magnetism so exhausts the magnetizer. Certainly it is not his exertions, for a few passes cannot fatigue. Few persons, however strong and robust, can magnetize to any great extent without feeling exhausted, and persons of feeble constitutions are extremely fatigued. The explanation is self-evident. The magnetizer imparts to the magnetized his own state of vibration, and to do so exerts his will so strongly as to exhaust its energy, and, as will is the parent of muscular force, of course he will become debilitated in proportion to the length and degree of his exertions.

This is still further proved by the effects of magnetism on the magnetized. When laboring under disease, magnetism invigorates the constitution, and in many cases works a radical cure. A case lately came under our observation. A consumptive, too feeble to move, and trembling on the very brink of the grave, arose under the zoethic influence, and walked about the room; yet, when this influence was expended, she was as weak as previously, and in a few days expired.

Though we are surrounded by such an atmosphere, we have no instruments by which to ascertain its presence, as we do that of electricity by the electrometer. The only reliable test is the impressibility of the brain. The brain *feels* its presence, and is to it what the most delicate electrometer is to electricity, or the finest iodized plate is to light.

There is an influence exerted by individuals unconsciously on each other which cannot be felt by the nerves in their ordinary state, but which is plainly seen by aid of clairvoyance. To the spiritual eye, every individual appears like a luminous centre throwing off zoethic waves in every direction, as a lamp throws off waves of light.

By the impressibility of the brain a new branch of mental science has been developed, of great interest and importance; and, as it illustrates the subject under consideration, we shall devote a chapter to its philosophy.

The word *psychometry*, by which the discoverer designated certain mysterious relations of mind, is derived from the Greek, and means, when translated, *soul-measure.* It is rightly named, for it measures the thoughts, and leads directly into the most secret recesses of mind.

This science depends on the impressibility of the brain — a faculty already proved to exist by numerous facts.

In making experiments in this department, or in any other relating to mind or spirit, the greatest care should be used, and the few necessary conditions already known complied with in as perfect a manner as possible. The student of the physical sciences deals with elements he can see, feel, and measure. He understands their properties — can combine them, and observe the result. If he places iron and sulphur in a retort, and applies heat, he knows that sulphuret of iron will be produced; and that he will obtain water if he burns oxygen and hydrogen gas together. In all these operations he can pronounce with certainty what the effects will be, for he can fulfil all the necessary conditions. Not so, however, with the student of psychological science. He enters a new and unexplored realm, and deals with elements so ethereal and subtile that they lose all the properties usually attributed to matter, and become more properly *agents* than elements. He cannot see nor measure them; nor can he fulfil the required conditions, for he does not know what they are. His steps are empirical, and the results obtained subject to great detractions. Suppose the student of chemistry could neither see, feel, nor measure the elements with which he experiments, and knew little of the laws by which they influence each other; how uncertain must be his tests! Yet such is the position of the student of mental science in regard to the elements with which he experiments. As he knows but little of his subject, he must rigidly comply with known conditions. This applies equally well to all physical research. Too great care cannot be used in observance of known conditions.

Psychometry depends on *nerve-aura*, or *zoether* — in fact, on the same law as that by which one person influences another. The animal pursuing its prey by its track, and the impressible individual revealing character by a garment or autograph, exercise a kindred faculty. An influence in all such cases has been left, which is felt by the brain. The peculiar state of vibration of one brain is induced in another. It may seem incredible that any influence whatever can be left on paper by simply writing a name on it, and still more incredible that character can be delineated from it. How this results we will now explain.

The brain is divided into groups of faculties. The zoether's vibrations from these are all unlike each other. Thus combativeness throws out a certain amount of zoether, destructiveness its amount, acquisitiveness its amount, and the intellectual and moral groups other portions. The combined aggregate of all these is the aura, or zoethic sphere, of the individual. Harmony, correspondence, every where prevails. If combativeness is large, it will stamp its influence on a large amount of zoether; if morality or intellect is large, it will originate a large proportional share of the zoethic sphere. Let us observe nature. The lion and tiger possess combativeness and destructiveness in an eminent degree. The contour of their bodies speaks this plainer than words. See the flowing mane, strong limbs, the prominent muscles! Hear their terrible roar and harsh growl, which send the affrighted quadrupeds flying over the plain! Do they not indicate a cruel, bloodthirsty disposition? If we turn to human nature, we shall soon meet cousins to these — men who are lions and tigers in every word and deed; with hoarse, harsh voices, and stern, unfeeling action.

The squirrel's prominent front teeth and sly motions, working all day to lay up his winter store, speak his large acquisitiveness. We often see men with just such teeth, and countenances contracted like the squirrel's, who not only work all day, but all night likewise, to lay up a useless hoard.

The man of intellect, how easy and lofty his bearing! His countenance is unruffled except by the great thoughts within, and those thoughts fashion the body.

This yielding of the body to the dictation of the mind, as though it were a plastic material, is easy of explanation; for before it forms the zoethic sphere around the individual, the vibrations of the organs of the brain pass through the body and impress it with their influence. If acquisitiveness is the largest or controlling organ, it throws off more than a due proportion of zoether, and detracts from all the others; and pervading the body, we find that it yields to its influence, and an acquisitive expression steals over the face, the hands are clinched, the step is cautious and infirm.

If one organ becomes excessively active, the tendency is to weaken all the others, which gives it still greater proportional strength, as it feeds on their food. Hence it is that they who commit excessive crimes, or are habitual drunkards or gormands, seldom reform, for the faculties commiting such acts have passed from the control of the moral, and are the controlling forces of character.

The size and activity of the various organs of the brain, and their individual proportions of zoether in the sphere of the brain, are intimately related, and one can be determined by the other.

When an individual writes on paper, his sphere imparts its state of vibration to the paper; and from what has been said before, it will be understood that that state represents all the faculties in their true relation, and hence, if analyzed, would give the size of the organs from which it originated, or the character of the writer.

Fortunately, the brain is the best of analysts, and by its impressibility the very thoughts of the writer, at the time of writing, can be determined, and his countenance and manners faithfully told. This is one of the most beautiful laws of nature. That an autograph, or scrap of writing, contains the active elements of the writer's character, and that in their relative proportions, seemingly belongs more to the dreamland of fancy than to philosophical research; yet rigid demonstration is easily obtained.

The result is the same if a lock of hair or fragment of wearing apparel be employed instead of an autograph.

It has long been the custom to send the clairvoyant physician a lock of the patient's hair, as it has been found that a lock of hair contains all the elements of the individual's character to whom it belongs.

As in magnetism his zoethic sphere reproduces the thoughts of the magnetizer in the magnetized, so in the autograph it reproduces the precise action of the brain by which it was produced, and consequently the same thoughts, more or less distinct, according to the impressibility of the psychometrist.

Not that the individual, while performing these experiments, is magnetized — no trace of this can be discovered; but it succeeds best with those who have the organ of impressibility large; and it must be admitted that the mind is influenced in precisely the same manner, though not in the same degree. The two are identical in nature, varying only in quantity. In one the whole energy of the mind is employed, while in the other only a scrap of writing can be used.

This is beautifully proved by an impressible person placing his hand on the head of one whose character he wishes to delineate, or taking hold of the hand; and the impressions thus received will arise much sooner than from an autograph, and be much more sharply defined, partaking more of the character of magnetism.

This is a very correct method of obtaining the character, and far excels phrenology; for, while the latter tells what the result of a given combination of organs should be, psychometry tells what they *are*. It enters the depths of the mind, lays bare all its thoughts and emotions, and, with a deep, penetrating gaze, understands the *man*. Hence it can give better counsel which faculties to control, which cultivate, and how to form a true and noble character.

Almost every one is susceptible to this influence. Not more than one in ten in the middle classes but might feel it in a lesser or greater degree. We say middle classes, because in the poor and lowly, who suffer from want and starvation, and the miseries of poverty, this faculty is rarely developed, and in the wealthy, circumstances are almost equally unfavorable. One in twenty-five is capable of excelling in impressibility. The organization determines this point. The good psychometrist may be known by a full, well-balanced brain and nervous temperament.

When an autograph or letter is taken in the hand, the sensation is first felt in the hand, gradually extending up the arm until it affects the mind. The same sensation is produced when it is placed on the forehead, but is experienced more rapidly. In either case the psychometrist should endeavor to remain as passive as possible — free from all excitement.

A word might be profitably said on the choice of autographs for tests. None are as good when long mixed with other writings, as the influences from the papers thus brought in contact blend. This occurs in the most remarkable manner when a very negative and positive letter are folded together. The character delineated from either will not be true, but each will be blended with the other.

The character of the person will be delineated as at the time of writing, and not its general features. This is philosophically true when the theory previously advanced is considered. If the letter is written under the influence of combativeness, acquisitiveness, or morality, of course these particular organs will be delineated larger than they in reality are. Allowance should always be made for this, and the science not condemned, for this is a convincing proof of its correctness. This source of inaccuracy can be avoided by using several letters written in different moods of mind.

So, too, the organization of the psychometrist affects the delineation. Thus, if he possess large and active ideality, and the writer of the autograph have it small, he will be very apt to delineate it larger than it should be. This error is of little account in the most impressible minds, but is very serious in those of a lower order of impressibility.

The reading of character is not the only application of this discovery. It is a good bark for the historian and antiquarian, carrying them up the stream of time — far beyond the confines of written records. How grand would be the true character of Alexander, Cæsar, or Napoleon, obtained in this manner, free from the prejudices of their biographers and their times! We are guided by fragments in our course. The linen which shrouds the mummies of Egypt reveals the character of that class who were considered worthy of embalming, and a fragment of Herculaneum will give Roman character two thousand years ago. The characters of the men who scattered mounds and fortifications over the American continent may be determined by their relics.

Nor does this all-penetrating science rest here. It takes the paleontologist by the hand, and leads him down through the carboniferous shales and sandstones, and by aid of the smallest fragment of organic remains gives him a perfect view of the world, at that age of its development. It revels amid the extinct beings of a former world.

Geologists have long sought to determine the real aspect of ancient nature; but, having no grounds on which to rest their speculations, of course they were only reckless efforts of the imagination. It is said the world was made for the use of man, and beautified to give him enjoyment. All the beauty and grandeur of previous times, however, is shut out from his gaze; but this science opens to view the vista of the ages.

In the detection of poisons, and the selection of

medicines, it points out their uses and abuses, and action on the mind and body.

Homœopathic doses, here, have as decided effect as allopathic, for it does not depend on quantity, but quality.

Such is psychometry in its relations to impressibility of the brain, and magnetism, of which it is but a modified application.

CHAPTER XIII.

ANIMAL MAGNETISM, ITS PHILOSOPHY, LAWS, APPLICATION, AND RELATION TO SPIRITUALISM.

Clairvoyance the Harbinger of the next State. — Incomprehensibility of Mind. — Mind can become independent of the Body. — Its six States: 1. Activity and Repose; 2. Impressible State; 3. Magnetic; 4. Clairvoyant; 5. Super-clairvoyant; 6. Death, or Independent-Spiritual. — Description and Illustration of these States. — Explanation of Impressibility. — One Mind can control another. — Philosophy of such Control. — Illustration. — Spirit Intercourse through Impressibility. — Its Difficulties. — Low Spirits (*Evil?*) — Their Habitation. — Influence. — Physical Manifestations, how produced. — By what Class of Spirits. — Spectral Apparitions, how produced. — One Law holds good in the entire Domains of Magnetism and Spiritualism. — Proofs and Illustrations.

From the natural state in which man normally exists, to the gateway of another sphere, where the silver cord which unites the mortal with the immortal, is broken, a wide interval exists. In the natural state the material has the ascendency, and the spiritual is subordinate. At death the spirit obtains complete ascendency, and the body fades. Between these extremes all degrees of separation exist. The two are variously blended, as light and darkness blend at morning; night representing the body, light the spiritual life, which slowly breaks on the horizon, gradually increasing, until the sun at length pours a flood of splendor above the gray clouds of morning. Then the spirit is free, and beholds the supernal light of the spheres.

By magnetism the phenomena of death are obtained, and its laws can be studied. It is then the right means to employ, as by it we reach, and as it were analyze, the spirit.

Spirit, and its essence, the mind, evade the scalpel of the dissector; it cannot be examined in the crucible or retort; it is unseen by the eye, unheard by the ear, and is only recognized by its effects. Yet it must be material, or the effect of materiality, for without an adequate cause there can be no effect.

The phenomena of physical agents cannot unlock its mysterious domain; and if any thing is learned of its nature, it must be by studying itself — not by the rushlight of metaphysics, but by the clear light of positive science.

However dependent it may appear to be on the body, there is an extensive range of facts which prove that under certain conditions it may become independent. When studied on the plane of physical science, it seems to have an exclusive dependence on the body, living while it lives, and dying when it dies. But there is a higher position from which to study mentality. It is unique, and must be studied by the light of itself. The recent discoveries in mental impressibility, clairvoyance, &c., open a wide avenue for the venturesome student to enter the halls of mind. The opportunity has been eagerly seized. Forsaking the beaten path of the metaphysician, the inquirers have pushed boldly out into the realm of facts and causes, and sought to construct theories in harmony with nature.

The observed facts of magnetism show that mind can in different degrees become independent of the physical body, and in proportion as it becomes independent does its spiritual perceptions become acute. This independence regards the senses and the entire organism, and the mind rises above the aid it furnishes, seeing, hearing, and feeling, independent of its organism. For classification of facts, the mind may be considered in six different states or degrees.

1. The natural state of activity and rest.
2. The impressible state.
3. Magnetic.
4. Clairvoyant.
5. Super-clairvoyant.
6. Death, or the independent spiritual condition.

1. *The Natural State of Activity and Rest.* — In this state the mind is chained to the body, and its manifestations are limited by the capacity of the latter. It sees with the eyes, hears with the ears, and feels through the agency of the sensatory nerves. To all appearances it is indissolubly connected with it, and from facts elicited from this state, the sceptic triumphantly exclaims, that it is as rational to look for the hum of the bee after the insect has passed on its busy wings, as for mind after the death of the body.

In this state there is a perfect union of the two, and their action is so blended that it is with extreme difficulty that the manifestations of one can be distinguished from those of the other. The mind never grasps any thing by intuition while in this state, but is content to plod in the grovelling externalisms of life, relying wholly on the five senses for its knowledge.

2. *The Impressible State or Degree.* — By this state we mean that condition in which the individual is susceptible to the influence of surrounding objects and minds. It is the normal condition of nearly one fourth of the Anglo-Saxon race. It varies in degree from the impressibility which shapes our attractions

and repulsions, to that which enters the secret cham bers of another's thoughts, and makes itself familiar with the innermost shadings of character. In the superior conditions of this state, psychometric delineations are made perfect according to the degree of impressibility and the peculiar influence of individuals become perceptible.

3. *The Magnetic State* is a higher degree of the last. It is not necessarily induced by an operator, instances occurring repeatedly where it has been entered spontaneously.

The mind is one step farther removed from the body, and now first manifests its independence by seeing, hearing, feeling, without the aid of the bodily organs, and reading the thoughts of those with whom it comes in contact.

It may be produced by disease.

Mrs. Sanby relates an instance of a young friend who had the regular functions of the nervous system overthrown by the sudden news of the death of her father. During the attacks peculiar to the disease thus induced, which is known to physicians as the Protean disorder, she possessed all the powers of the true somnambulist. The extraordinary powers communicated to the other senses by the temporary suspension of one or two of them, are beyond credibility to those who do not witness it; all colors she can distinguish with the greatest correctness by night or by day; . . . and I may safely say as well on any part of her body as with her hands. She can not only read with the greatest rapidity any writing that is legible to us, music, &c., with the mere passing of her fingers over it, whether in a dark or light room, (for her sight is for the most part suspended under the paroxysm,) but she can read any book or writing by simply placing her hand on the page. Such facts not only prove the possible independence of the spirit from the body, but also that it acquires a sense superior to the five bodily senses.

This state can be produced by an operator whose positive sphere blends with and overrules the negative sphere of the impressible subject.

4. *Clairvoyance.* — The supremacy of one mind over another is best seen in clairvoyance, in which mind rises entirely above the corporeal senses, sees objects at the remotest distance, hears and feels independent of the body, and of all physical organisms.

5. *The Super-clairvoyant* is a state of independent clairvoyance — another step upwards — in which the spirit leaves the body, and, united with it only by the finest cord, traverses the remotest regions, converses with superior intelligences, and after its wanderings again returns to the physical body. During the continuance of this state the body is motionless, vitality almost ceases, the blood scarcely flows in the veins, the lungs scarcely inflate, and all the other appearances are those of death.

[I cannot refrain from introducing here an incident of my own clairvoyant experience. It occurred before I was well acquainted with the *modus operandi* of the process, and hence the anguish I experienced. I apparently left the body, and in company with my guardian, went to the spirit world. I was ecstatically happy, and, seemingly with my senses a thousand-fold augmented, took cognizance of every thing around me; but rapidly fled the hours, and I knew I must return. I came to my body. I saw it cold and motionless, rigid in every muscle and fibre. I endeavored to regain possession of it several times, yet could not, and by my successive failures became so alarmed that I could not even make the effort; and it was only through and by the influence of the friends who were present that I succeeded at all. When at length I did recover my mortal garb, the anguish, the pain, the agony of that moment was indescribable. It was like that which is used to describe death, or which drowned men tell us of when they at length recover.

The passage from the body was pleasant, agreeable, joyful. A current of fresh wine coursed through my veins, but the return thrills with agony when I recall it. I was outside of my body. I looked on it as any other foreign substance. It was all reality to me. Now, how can it be that the clairvoyant does not leave the physical body? Some assert that such is not the fact. If so, then good by to clairvoyance forever; for its teachings are too vague for embodiment in a system. All its revelations stand on the same platform, and if one is discarded so must all.]

6. *Death, or the Independent Spiritual State.*— When the thin cord which unites the spirit with the body is broken, death results. Then the spirit cannot again enter its earthly tabernacle. The vital principle which animates its mechanism has fled, and, as a useless garment, it moulders back to earth.

How closely super-clairvoyance approaches death, is seen in Cahagnet's seance with his ecstatic Adélé She had repeatedly assured him that there was great danger in her ecstasy, as it might be carried too far, and her spirit be completely severed from her body.

He wished to satisfy himself on this point, and allowed her to sink as deeply into that state as she pleased, having another clairvoyant to watch her, and give the alarm should any thing serious occur. At length the latter exclaimed that he had lost sight of her. On examination, not the slightest pulse was discernible; and on holding a mirror to her lips it was not tarnished. He magnetized her with the most powerful efforts of his will, but for a long time could not produce the slightest effects. Her first words were, "Why have you called me back? It was all over with me; but God, moved at your prayer, sent me back to you. No more shall I be permitted to return to heaven. I am punished. . . I shall no longer be able to ascend to heaven; but had it not been for you, I should have been there now, and forever."

Death opens the portals to the next sphere, and the spirit always sees its spirit friends before it departs. In illustration of both the foregoing positions, we introduce the following facts: —

It has been remarked that when a person faints he loses all remembrance, and passes into a state of total forgetfulness; but if the syncope be prolonged to the verge of death, and the person then recover, he will remember things that occurred in that state. In other words, the memory is a blank in the first stages of mesmerism, but in the higher stages, bordering on death, it is active. The relation of instances to this point will illustrate.

A lady departed this life under an influence which caused repeated fainting.* When she recovered from

* Cause and Cure of Infidelity, by David Nelson, p. 264.

the first condition of syncope, she appeared unconscious of what had transpired. She sunk again, and revived; it was still the same. She fainted still more profoundly, yet, when she revived, could not recall her thoughts. At length she seemed entirely gone, and her friends thought the struggle was passed; but she revived, this time fully conscious of what she saw in the trance state, which now began to dawn, as the spiritual rose above the material, and clasping her hands together, exclaimed, " Ah, I was in an entirely new place!" and fell back, this time, into the embrace of death, which transported her immortal spirit to that beautiful place she saw in her previous trance, while the cast-off body remained to moulder back to dust, and enter new organic forms.

Dr. Rush records an instance where a man supposed dead recovered. While in this trance, his mind was extraordinarily active, and he heard and saw unutter able things. St. Paul, in a similar state, could not tell whether he was in the body or out of it, nor could he describe by words what he saw.

A case is recorded * of a revolutionary officer, who, on his death bed, made an agreement with his daughter, that when the new world revealed itself to him he would press her hand. He was a good and truthful man, yet his mind was clouded with doubts, and he feared to approach the dark river. Hour after hour she sat holding his hand, waiting the promised signal. The struggle had passed. He lay still and passive, drawing the slow, gurgling breath which proclaims dissolution. One by one his senses were closed; his

* Cause and Cure of Infidelity.

vision failed, his hearing, his speech; yet life remained. Still the spirit was not sufficiently free to see its future abode. At length that super-clairvoyant stage was reached; he pressed his daughter's hand; a gleam of light radiated his countenance; and that moment the breath ceased, and his spirit soared to its immortal home.

The propositions which we advance and seek to maintain by the facts classified under the foregoing divisions are, *Mind can exist independent of the physical body; Mind is referable to the spiritual body.* In their support we bring forward the vast volumes of facts of prevision, prophecy, clairvoyance, and magnetism — a force sufficient to crush all opposition. It is useless to fill our pages with a detail of facts with which every one the least versed in psychological science is familiar. Suffice it to glance at the condition of the clairvoyant subject, and remark the exact relations which exist between the mind and body.

The subject has entered the impressible or clairvoyant state. Slowly the vital powers sink until the body becomes as it were dead. It is insensible to pain. Even the excito-motor nerves produce not the least motion. When lacerated or burned there is neither sensation nor movement. The mind can see and hear at an immense distance, or else it leaves the body and traverses the regions of space. It reads the thoughts of persons at a great distance, and when it comes back retains a vivid remembrance of the strange scenes it has witnessed, and *testifies to the fact that it was really detached, except by a slender connection, from the body.*

The evidence is of persons who have passed through

the clairvoyant portals. Can more positive testimony be desired? From some cause this subject has become involved in doubt, and its statements received with puerile incredulity. We must not enter the field of mental science with the diagram of the geometrician, or the crucible and retort of the chemist. These are well in their places, but on entering a new study they should be laid aside, and the facts furnished by mind itself alone receive attention.

From the normal state to the death of the body, or complete separation from it of the spirit, we see successive steps by which the mind leaves the physical form. In the normal state it is inseparably united; then it rises partially above it, and manifests the newly-acquired faculty of impressibility; then it becomes magnetic and clairvoyant, and exhibits a noble freedom from corporeal restraints. It sees when the eyes are closed, hears the slightest sound when the ears are tightly sealed, and by its superior knowledge conclusively shows that the body is rather detrimental than auxiliary to the expansion of thought.

In the normal state there is a mutual dependence of the mind and body which qualifies man for the earthly sphere. Born in intimate relations, nourished together, supported by the aid they furnish each other, there is of necessity a remarkable dependence. But on the part of the mind this is only seeming, not real. Back of nerves and brain, of cell and cell-contents, there is a necessity for higher and superior conditions —just as beneath all the changing phenomena of external nature great and incontrovertible principles are seen upholding on their Atlas shoulders all created things. We must go farther than external contemplation to account for the phenomena observed. We cannot refer mind entirely to the body. It does not originate in the changes we saw transpiring in the brain; these were means of its manifestation; and, when the complicated, nervous structure was described, it was considered as the engine without steam, nicely adjusted for the operations of intelligence, but inert until moved by that superior force derived from the spiritual aggregate of refined matter which composes the spirit body.

If mind was wholly dependent on the physical body it could not act without it. Clairvoyance would be as impossible for man as for brutes. But clairvoyance is established, and yields a weighty argument — rather invincible demonstration — that mind can become independent.

If, in clairvoyance, all means of deception are destroyed, and the subject retains all the senses unimpaired, although the external organs are sealed, then the independent existence of mind is demonstrated. Not only one case, but innumerable ones have and are occurring of the strictest independent prevision and spiritual sight.

If the decline of the intellect, in old age, is brought forward in support of the dependence of mind on the body, the counter-fact can be arrayed against the conclusions deduced therefrom.

There are men who, like Humboldt, to their oldest age retain their intellectual powers unimpaired, and, like him, can note the decay of the physical form, mark each change, and calculate with the calm eye of philosophy the period of dissolution. He devoted life exclusively to the cultivation of his intellect, and advanced beyond the influence of physical decay.

While his body was ready to fall into the grave, his spirit was unimpaired, and ready to become an independent being so soon as the thread which binds it was broken.

Mind is an effect of superior causes, and if those causes do not reside in the physical form, there must be some higher source to which it is referable. Back of the external phenomena must be a hidden power; that power is the spiritual nature of man, as incarnated in his spiritual body, to which mind must be referred as an effect.

To investigate this realm we must light the torch of mesmeric revelations, and enter at once its intricate pathway. We are to search for the explanation of the phenomena described in the foregoing pages, and show how mind can influence mind, and how enter the impressible states.

We have written of magnetism, its different states, and corresponding phenomena. We may now inquire what is the prime cause of all these results. It is certain that there must be a medium of communication, otherwise no influence could pass from one individual to another. Even intangible motion cannot be communicated without the intervention of tangible matter. If one individual influences the thoughts and actions of another in a distant apartment, simply by the effect of his will, then it is self-evident that something passes from one to the other. This proposition does not require proof; it is as self-evident as the axiom — nothing cannot create something.

What is this something? Facts conflict with the hypothesis of its being matter radiated from one individual to another, as light was once supposed to be transmitted. It darts with too great precision, is too instantaneous in its action, passes too readily through vast thicknesses of solid matter to consist of radiant particles. On the other hand, all these phenomena show a striking relationship to light, heat, and kindred agents, and whatever explains one, is alike applicable to all. We have already discussed this subject, and to the universal ether-ocean referred these phenomena, and as waves in this medium, of a certain length, produce light, of another length heat, of another magnetism, so of another length they produce psychological phenomena. As a luminous body is capable of producing waves of light, a living being is capable of producing zoethic waves. These waves are transmitted with greater rapidity than vibrations of light, their velocity being about 250,000 miles in a second.

Now, let us inquire how, by means of these undulations, one individual can influence another.

According to the above theory, the brain vibrates like the strings of a musical instrument; and as no two brains are exactly alike, so no two vibrate alike. This illustration is more than merely an illustration. Both depend on similar laws, for the string excites vibrations in the air which are felt by the tympanum of the ear; the brain excites undulations in ether which are impressed on other brains. The nervous system alone can feel these waves. The string of the instrument excites similar vibrations in contiguous strings; for the atmosphere transmits the waves of sound, or being set in motion by one string, by its momentum sets the other string in vibration.

This is very beautifully shown by a simple experiment, which equally well illustrates the method by which mind influences mind.

If a plate of glass is strewn with sand, and, while held in a horizontal position, a bow be drawn across its edge, a musical sound will be produced from the vibration of the plate, and the sand, by the vibration, will be thrown into various geometric lines, according to the note produced — each note giving rise to a figure peculiar to itself. So invariably is this observed, that a piece of music might be accurately written from the forms assumed by the sand.

Now, if a piece of parchment or paper be stretched, with proper precautions, across the top of a large bell glass and strewn with sand, and the glass plate be held horizontally over it, and the bow drawn across its edge, the forms assumed by the sand on the paper will accurately correspond with the forms on the glass. If the plate is slowly removed to greater and greater distances, the same correspondence will exist, until the distance is too great for the air to transmit the vibrations.

If the plate, while vibrating, is held perpendicularly to the horizon, the figures on the paper will form into straight lines parallel to the surface of the plate, by creeping along it, instead of dancing up and down. If the plate be made to turn around on its vertical diameter while vibrating, the lines on the paper will revolve, exactly following the motions of the plate.

When a slow air is played on a flute near this apparatus, each note calls up a particular form in the sand, which the next note effaces to establish its own. The motion of the sand will even detect sounds that are inaudible. Besieged armies have discovered the direction in which the countermine was working, by the vibration of sand on a drumhead.

Professor Wheatstone has beautifully illustrated this correspondence, or rather *sympathy*. If a sounding-board is placed so as to resound to all the instruments of an orchestra, and connected, by a metallic rod of considerable length, with the sounding-board of a harp or piano, the latter will respond to the exact notes of the former.

The effect of this experiment is very pleasant; the sounds, indeed, have so little intensity as to be scarcely heard at a distance from the reciprocating instrument; but on placing the ear close to it, a diminutive band is heard, in which all the instruments preserve their distinctive qualities, and the pianos and fortes, the crescendos, and diminuendos, their relative contrasts.

The nervous system is inconceivably finer organized than the most perfect instrument, and if such delicate sympathy is exhibited by the latter, how much more perfect must we expect it in the former! The nerves, like tense strings, feel the slightest vibration in ether. The brain records each vibration so received. Such is the general statement, which, however, must be qualified; for it teaches that all minds can influence each other, which is not true.

Light falls on all substances alike, but is very differently affected. One class of bodies absorb all except the yellow rays; another all but the blue; another all but the red. Why is this? Because these substances are so organized that they respond (sympathize) only to waves of these colors.

Some individuals have the ear so organized that

they can hear certain sounds, but are totally deaf to others. The waves of sound strike all tympanii alike; yet in these instances they are incapable of responding to certain waves. Some persons, who delight in music, although all the lower notes are plainly heard, as soon as the tune rises to a high key, cannot hear a single sound. In others this is reversed; the high notes are audible, but the low ones are lost. The eye of some individuals is similarly arranged, some colors being undiscernible, while others are perceptible. For the cause of such effects we are not so much to examine the ether as the construction of the nerves. We know that the same vibrations exist in all instances, but owing to peculiarities of organization many of them are not felt. The law is manifest in the experiment with two musical strings stretched parallel to each other, one being twice or three times the length of the other. If the shorter is set in vibration, the longer will divide into two or three segments, and vibrate in the same rate. In order to have one string cause vibrations in another, there must exist some such relations. There must be a similarity. So, for one brain to transmit its vibrations to another brain, a similarity must exist. As a musical string extended over a bar of wood or iron cannot transmit its state of vibration, so two brains entirely differently organized cannot respond to the vibrations of each other. All brains throw out vibrations, as all strings when extended give off waves of sound; but as the string must have a corresponding string to receive its vibrations, so the brain must have an harmoniously tuned brain to receive its vibrations.

Two waves of light (chromether), in opposite states of vibration, meeting destroy each other, and produce darkness; but if corresponding, a double luminosity results, as one is added to the other. The same is true of zoether.

In the accompanying diagram this important proposition is represented to the eye.

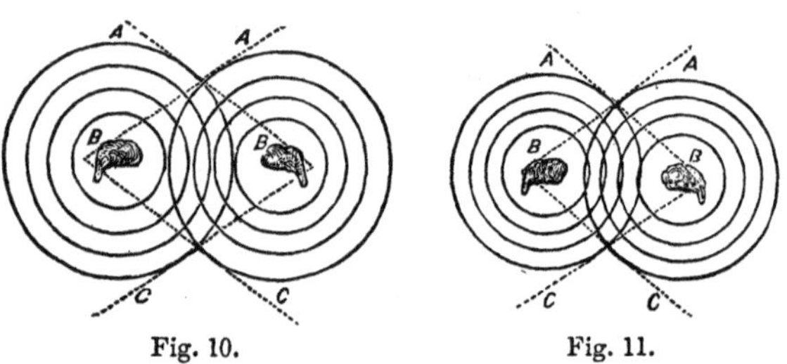

Fig. 10. Fig. 11.

Fig. 10 represents the vibrations of two harmonious brains, B B, and shows the manner in which they mingle. From them, as from two centres of luminosity, undulations go out; and if in harmony, each responds to the other. Fig. 11 represents two brains not in harmony, hence not affected by each other, as the waves impinge against and mutually destroy each other at their points of contact. It will be readily seen that only the waves between the lines A C can interfere with each other, and if allowed to pass they will impinge on the neighboring brain.

Fig. 12. Fig. 13.

Figs. 12 and 13 represent the same in detail. Fig. 12 shows the relation of harmonious vibrations, the hollows of one corresponding to the hollows of the other. Fig. 13 represents the relation of inharmonious vibrations. It will be readily seen that as these vibrations in Fig. 12 pursue the course from A to M they will do so uninterruptedly, but in Fig. 13 they will meet at the points L L L, and mutually destroy each other.

Here we arrive at the philosophy of all psychological influence, whether received under the name of magnetism, mental influence, or spiritual impression. One law underlies and ramifies through all these diversified effects.

When two individuals come in contact, if not harmoniously organized, at least in some point, they do not exert a mental influence on each other; but if, as previously shown, the necessary conditions of organization are complied with, they will, in a greater or lesser degree, exert such influence on each other.

This is unavoidable, whether the will is exerted or not; but if the stronger will is exerted, its power is proportionally greater, and it will subdue and control, that is, mesmerize, the weaker, and the peculiar phenomena arising from one person having his will controlled by another will result.

We thus perceive how one mind can control another. We have proved that such control is possible, and have shown how it is effected.

So far we have considered the relations of mind as connected with the body. Now let us turn to the relation these facts and theorizings bear to spiritual intercourse.

It is not the body which magnetizes or is magnetized, it is the mind; and it produces these effects outside of the physical system. In proof we need but adduce the fact that one person can magnetize another by the simple power of will, though the two are a thousand miles asunder. Here we see unaided mind producing the most startling zoethic phenomena, and as it were detached from the physical body.

A mind thus situated is in the same position as a spirit. It is freed from the physical, except that it has greater freedom, and is more exquisitely susceptible to the influence of other minds. Hence it will be readily perceived that there is not the least obstacle in the way of one spirit impressing his thoughts on another harmonious spirit. The same law holds between them as between magnetizer and magnetized. And as man is a spirit incarcerated in a body, and in that respect only differing from a disembodied spirit, his want of susceptibility alone debars him from intercourse with spirits above him.

You think it strange that so few can hold this intercourse, but is it not equally strange that so few are capable of passing into the mesmeric state? Not one in a thousand possesses this faculty sufficiently to read the *thoughts* of others, yet it must be possessed in an eminent degree to read sentence after sentence from our minds. How much more susceptible must the mind become to write a volume, word by word, even should numerous mistakes occur!

Yet men say we err — that our communications are false. Ah, they think not that the stream is shaped by the channel, on which depends whether it be a clear mountain torrent, agitated by falls and rapids, or a placid stream; whether it be crystal, or rank with the miasm of swamp lands. A given aperture can only permit a certain amount of water to pass. The medium's mind may be impressible with some thoughts, yet wholly unconscious of others, as in the instance given, the ear may hear certain sounds, yet not others; or the eye recognize certain colors, yet be unconscious of others. Let us suppose we are endeavoring to communicate through a medium thus imperfectly organized; that his animal propensities are very strong, while his morality is very weak. We endeavor to communicate on morals, and may occasionally succeed in impressing a single idea so vividly that it will remain unmixed with his own mind; but as a whole the communication will be tinged with his predominant animality, and however pure our thoughts, they will flow from him mixed with the soil of passion.

Suppose you who doubt, magnetize the most impressible person you can find, and when you think him completely under your control, endeavor to make him speak what you think; and how many mistakes will occur! If the subject be very good the ideas will be given, but the wording will be incorrect.

The following instance illustrates. The subject was extremely susceptible. It is reported by Capron, in Deleuze. A sealed letter was given the magnetized, who read it, —

"No other than the eye of Omnipotence can read this in this envelope.

"———————— 1837."

The true reading was, —

"No other than the eye of Omnipotence can read this sentence in this envelope.

"Troy, New York, August, 1837."

Many may refer this and kindred facts to spiritual intercourse rather than a law of mind; but as we have before advocated, we had rather meet the subject fairly, and explain the cause of phenomena, rather than refer such as we cannot account for to spirits, for the sole reason of our own inability to explain them. Such facts as are really spiritual we shall refer to that source, but such as are not we shall expound by other means.

Such are the difficulties of spirit intercourse. The magnetizer will appreciate them when he endeavors to impress his thoughts on his subject. But he *can* do so, and so can we, and often with complete success. But the sources of error are numerous, the channel imperfect, and hence sentences will flow from the medium widely different from those we strive to utter. The higher, purer, and more intellectual the spirit and the medium, other things being equal, the purer the communication; and the lower and more debased the spirit and the medium, the more inaccurate and distorted the ideas which will be given. The more elevated the spirit, the better it can control mind; the lower the spirit, the better can it control matter.

Another form of impression somewhat differs from true magnetism in being mechanically effected, the medium's hand being moved to write sentences foreign to his mind, or in an unknown language. We do this through the medium of the mind; but consciousness, so far as our operation is concerned, is negatived by our will acting only on the nerve-ganglion which propels the arm, and hence the influence is not transmitted to the brain. This will explain all the complicated automatic movements which are not controlled by the will of the medium; the ganglionic system being disturbed by guardian spirits for the medium's benefit, or by low spirits for amusement or revenge. But it may be objected that the medium is often impressed without being in the least magnetized. This is pronounced without due reflection, for in so recent a science little positively can be known. But admit this to be so — are not individuals often impressed by others when apparently in a normal state? Certainly; but the mind in such instances, although it may not indicate it, is in a similar state to the mind of the subject when magnetized. The necessary condition in both cases is a high degree of susceptibility.

We say that undeveloped spirits have greater control over physical matter than spirits more refined. If we observe the conditions of their being we shall see that this must be so. We shall see that their home is on the surface of the earth, and that the objects which excited their desires while on earth still surround them. Their aspirations do not rise above things of earth, and they are content with their position. In many cases rising from the reeking carcass of sensuality, bestiality, drunkenness, gluttony, and crime, the taint of earth still clings to their garments. Hence they yet possess the brute strength of earth; and when elevated spirits desire to manifest physically they employ these as agents to do their work in their name.

How do they move matter? It is plain not by hands, as man does. Matter cannot be moved by the application of spirit bone and muscle; but the spirit can control the imponderable agents already described, and through their instrumentality affect matter. These agents in nature control all gross matter beneath them. If you doubt their power, see the chain lightning shiver the centennial oak, and strew the plain with the fragments; hear the braying thunder; see the storm dash on with the force of an avalanche; more, look up into the starry vault of heaven and behold the fleets of worlds floated by the breath of magnetism, holding the universe in the palm of its Atlas hand! Ah, now do you doubt that such forces, if controlled by intellect, can move a table or other furniture across a room?

Zoether emanating from the medium charges the object to be moved, and a band of spirits directs a current of their own zoethic emanation in the direction they desire the article to move, and it passes along the current thus produced. The charging of the object by the medium is necessary in order that the object may be in a state of vibration harmonious to the spirit current. If this current be directed against the table or other charged body, raps or concussions are produced, as a positive and negative relation exists between the spirits' zoether and the medium's.

We would here correct a mistaken idea which is prevalent. One spirit alone cannot communicate in the latter manner; i. e. cannot produce physical manifestations. If one purports to communicate, it is assisted by many others, who combine their influence as the members of a circle do theirs. In producing physical manifestations the spirits are with you — close by your side; but when they speak by impression they are not necessarily so. As one person can impress his thoughts on another at an indefinite distance; as the sister was conscious of the danger of her twin brother; as the sense of calamity fills the mind of one friend when an intimate is in danger,—so one spirit can influence another spirit, but with far greater certainty, and a spirit can influence the mind of a susceptible medium. Perhaps the spirit may be miles away, or even in the second sphere; yet his winged thoughts will come with the speed of light, and vibrate on the susceptible brain.

Yet another class of phenomena present themselves which we refer to the magnetic laws. Spirits appear to the clairvoyant eye. Usually they present themselves wrapped in a flowing mantle of gauze, and appear as though enveloped in a cloud; but for purposes of tests they seem to appear in the garments they wore while on earth. No fact can be produced substantiating more positively the mesmeric character of this phenomenon. That the spirits that so appear are really dressed as they represent themselves is not for a moment to be entertained. It is a psychological effect, belonging to that class of facts in which the mesmeric subject sees whatever object the operator desires that he should. You may say again, Ah, you depart from spiritual agency too often. Nay, pause and consider. The spirit is not protean. It cannot by desiring, as has been very erroneously taught, assume whatever dress it pleases. All is reality here, stern and unyielding as with you. If the spirit desires to appear dressed as on earth, of a similar age, &c., as test, it forms the desired image in the clairvoyant medium's mind, who, deceived by the illusion, thinks he sees the reality.

There is nothing loose here—no chance for deception; and the phenomena have equal weight as though a real image conveyed the impression. There must be something outside of and above the medium — an entity; and that entity by these very means identifies itself.

But if a spirit manifests its thoughts in this manner, why not a conniving individual employ the same medium for purposes of base fraud and deception? We admit this as possible; but it is not possible for a magnetizer to identify by incontrovertible tests the presence of a spirit, nor to answer questions correctly which none but the spirit purporting to communicate can understand. When Laplace solves a problem which none but Laplace can propound, as you have witnessed, the idea of such a result being obtained from the low intelligence of a person capable of deception is preposterous.

We say one law governs both; yet the causes are entirely different, and should not be confounded.

We will illustrate what we have said, to avoid all misunderstanding. Suppose two individuals standing in relation of husband and wife. The wife is susceptible to the thoughts of her husband. When partially magnetized she can read all his thoughts, or he can will her to think whatever he pleases. He holds spiritual communion with her. She is his medium. Now, if the husband dies, his spirit would hold the same relation to her as before. In either state, in the body or out, he is a spirit, and so is she. The body is no barrier to their communion. His mind is still harmonious with hers, and his thoughts vibrate on her brain. He now, as a spirit, can control her mind. Perhaps she may not recognize his presence, nor even the source of her ideas, for they flow so readily through the brain, consciousness is deceived, and mistakes the vibrations thus awakened as spontaneous. The spirit husband, bound by the silken cords of love, reposes by the side of his mate, and in hours of slumber fills her mind with dreams in which much of the actual is shadowed forth. Again she thinks that he is with her, but awakes to the stern reality. Child of mortality, he has been with you, and is with you still. When danger threatens you he will protect; when you are in need he will counsel. Such, O man, is our inter-relation, and such the laws by which we imperfectly communicate with you.

CHAPTER XIII.

PHILOSOPHY OF CHANGE AND DEATH.

Wonders of Change. — An Arabian Fable. — Cycle of Organic Forms. — Cause of Change in the Universe.

[I WRITE the words I receive by impression, or I relate the clairvoyant scenes which pass before my spiritual vision — strange and unique ofttimes, yet irrepressible. The giant spirits which overshadow me, — how feeble the instrument they have chosen! how I gasp when attempting to utter their sublime wisdom; how weak are words to thought-lightning which illumes the darkest recesses in the millionth of a moment! alas to elude all attempts to describe! I transcribe their words as well as may be, hoping for better times.]

Wonderful is change. The Aladin lamp, ever producing startling apparitions, ever overturning and revolutionizing the physical and spiritual worlds! It is a mighty word, synonymous with progress. The human heart yearns for advancement; and only through this gateway can it go on. Well and grandly has many a writer spoken of mutation, metamorphosis — the ebb and flow of existence. Of it an Arabian writer has told a beautiful story, containing a deep philosophical truth. "I passed one day," says he, "by a very ancient and populous city, and asked one of its inhabitants how long it had been founded. 'It is indeed a mighty city,' replied he; 'we know not how long it has existed; and our ancestors on this subject were as ignorant as ourselves.' Five centuries afterwards, as I passed by the same place, I could not perceive the least vestige of a city. I demanded of a peasant, who was gathering herbs on its former site, how long it had been destroyed. 'In sooth a strange question,' replied he; 'the ground here has never been different from what you now behold it.' 'Was there not of old,' said I, 'a splendid city here?' 'Never,' answered he, 'so far as we have seen; and never did our fathers speak to us of any such.' On my return there five hundred years afterwards, I found the sea in the same place, and on its shores were a party of fishermen, of whom I inquired how long the land had been covered by the water. 'Is this a question,' said they, 'for a man like you? This spot has always been what it is now.' I again returned five hundred years afterwards, and the sea had disappeared. I inquired of a man, who stood alone upon the spot, how long ago this change had taken place, and he gave me the same answer I had received before. Lastly, on coming back, after an equal lapse of time, I found there a flourishing city, more populous and more rich in beautiful buildings than the city I had seen the first time; and when I would fain have informed myself concerning its origin, the inhabitants answered me, 'Its rise is lost in remote antiquity. We are ignorant how long it has existed; and our fathers were, on this subject, as ignorant as ourselves.'

Such is the perpetual revolution and unrest of the world. Where the miasmatic marsh putrefies in the sun to-day, to-morrow the rank flags and loathsome reptiles give place to populous streets and splendid edifices. Where the rude canoe battles the stream, to-morrow its pride shall be conquered by the broad-chested steamer. The ocean swallows up its coasts in one place, to vomit them in another. Its bed, like the firm land, is unstable. One jar of the earthquake, and they rise or fall. Mountains jut up, around which clouds nestle or valleys gape, in whose depths reigns eternal night. Continents rear their broad backs from the sea, or disappear. Oceans are formed or converted into dry land.

The globe, internally and externally, is replete with change. All beneath is a fluctuating sea of fire, spouting through volcanic vents, or, when confined, rocking the unstable crust. Around us spread oceans with ceaseless tidal flow, and above us expands an atmosphere which never rests.

The stars of heaven sway hither and thither. The moon periodically grows and fades. The year grows old and dies. Ever is it putting on different robes. In the spring we welcome a joyous maiden with blushing cheeks and sunny curls, blue laughing eyes, dropping with tears of happiness. She is crowned with flowers; her breath is redolent with the sweetness of clover; the increase of flocks and herds are her retinue; the young birds warbling joyfully, and the bleating of lambkins, her musical voice. In summer the maid matures and throws aside her flowers. She labors in the field, the orchard, the vineyard. By autumn she has become a dignified matron, with a garb of russet brown; her flower wreath is exchanged for one of heads of yellow grain and mingled fruits, and she gathers into her lap the sheaves of the harvest, the ears of the golden corn, the luscious apple, the peach, and grape.

But ere she enjoys the fruits of her toil, she becomes a maniac, and dressed in fantastic garb, hides herself in October haze, and finishes by killing the flowers and disrobing the trees, which she has labored to feed and clothe. The frosts wrinkle her features; the cold winds bend her form; she totters onward a while a decrepit hag, clad in ragged mourning weeds; totters on through the snow, with the wild winds tossing her silvery hair, and the enraged trees shrieking above her head; totters on through the storm, over frozen streams wrinkled with congealed agony, contorted like serpents frozen, by ice-sheeted lakes, over which the wind raves; totters on by cottages frozen without, ice-bound, icicle-eaved, but warm-hearthed and warm-hearted within; stops not, but on in the pine forest, where the snow-laden trees lost their green tresses, falls down and expires. To-morrow another blooming maiden comes with cheeks as red and breath as fragrant as the parent whose bones bleach in the forest. So, ever phœnix-like, youth springs from the ashes of decay. The fallen tree moulders back to dust, and is absorbed by other trees, by insects burrowing in its structure, by animals cropping the green grass growing above it.

The atom which existed yesterday in the ear of corn, to-day becomes assimilated in the animal, to-morrow may become a part of man, and thereafter originate an idea, which, incarnated, may overturn empires and states.

An animal dies. How rapidly its component particles disperse, and at once feed and nourish a myriad forms! The winds bear them on swift wings from the Atlantic to the Pacific. The rains wash them down to the roots of the forest trees; they are absorbed. The oak is strengthened; its massive column acquires new layers of fibre; its leaves a greener hue; its acorns are larger; the pine is mantled in a darker robe; its perfume is denser; the flowers deeper blush in their broad corollas. All the vegetable world rejoices, from the violet nestling in the bogs of the meadow to the tall fir high up on the snow-capped mountain. All receive a part from the generous winds — from the gigantic balboas of the African forest, colossal in dimension five thousand years before the flood, and the grand Washingtonian pines nestling among the Sierras of California, to the purple gentian by the wayside stream, and the carpeting grass of the meadow. Through the vegetable, the particles return again to the animal. The ox, grazing on the pasture, consumes former generations of oxen and of mankind. Death throws back the worn particles to greedy beings. Not a hair falls from our heads but is eagerly devoured. Moths and millers, bugs of all sizes and hues, fill up the catalogue of consumers. Whole forests of mould grow in a night on the wasted drop of paste. Innumerable swarms of beings sport in every drop of fermenting liquor — called into existence by decay.

Herbivorous animals consume the vegetable world. They feed the carnivorous tribes, which yield back again to the plant its long wandering particles. Every being is a beast of prey; or, as the poet has humorously expressed it, —

> "Fleas have little fleas,
> And these have fleas to bite them;
> And these have other fleas,
> And so ad infinitum."

Shakspeare, in more lofty strain, chronicles the same, when his Ariel sings, —

> "Nothing of him that doth fade,
> But doth suffer a sea change,
> Into something new and strange."

This ceaseless change was early recognized, and mystified human observation. Early were observers divided into two great classes; one denying the possibility of any mutability in that which is; the other saw nature in the light of eternal mutation and variety. So far did the latter carry this philosophy, that they denied the existence of any thing steadfast or enduring in the physical, moral, or spiritual worlds. "Every thing," said they, "is perpetually changing and revolving like the water of a river." So early was the reflective mind impressed with the endless permutation of forms — the cycle of ceaseless decay and renovation. Creation appeared like an ocean of restless tidal motion, ever changing, yet ever the same.

The first effort of mind, in its rude savagery, to account for phenomena, was by referring them, as effects, to superior intelligences. Hence sprang the innumerable allegories which have given romance and poetry their attraction. Like children, the ancients viewed nature perceptively. They saw no connection between cause and effect. Arbitrary rule of superior beings alone accounted for the results they saw transpiring. Of these allegories, of which each race and nation has its own, those of the fire worshippers is most massive and sublime, and the most truthful. The great fountain is Orsmund, from whom every thing flowed by a simple effort of thought; and into whom, after completing a cycle of three hundred and sixty thousand years, every thing will return. He is the fire, the all-vivifying soul of the universe; the vast ocean in which, like tides, the waves of existence and non-existence flow. Such is their colossal doctrine of change; from his bosom all things originally flowed; back to his bosom all things return to flow out again, purer and more refined. Such is the law of change. Motion is the purifying fire through which purer and more refined spheres of being are attained.

> "What though the sea with waves continuall
> Doe eate the earth, it is no more at all;
> Ne is the earth the lesse, or loseth aught:
> For what so-ever from one place doth fall
> Is with the tyde unto another brought:
> For there is nothing lost, that may be found if sought.
> Likewise the earth is not augmented more
> By all that dying into it doe fade;
> For of the earth they formed were of yore."

Fade to be resurrected in higher forms. Motion is the cause of change; and it is self-evident that the law of motion is not a circle, but a spiral of eternal progress. Every wave which breaks on the shore of time rises higher than its predecessor.

For corroboration, glance at the history of mankind, or, diving still deeper, of the globe itself. Every where change is equivalent to progress. Revolution after revolution has broken over the moral and political worlds. Revolution after revolution has shaken our physical globe. Great retrogressions have broken the continuity of advancement, and broad chasms have intervened; yet onward has rolled the car of progress, bearing the great world and its beings into eternity.

The cause of change is motion; and of progress, the refinement of the material on which motion is exerted. The force ever remains unchanged; but the objective matter on which it is exerted constantly varies, becoming perfected, and susceptible of more refined combinations. If not so, why were the earth and living beings, in the geological ages of the past, so imperfect, and why so uniformly advanced? To the great force of gravitation, living and dead, gross or refined, matter is alike; and hence the rude vapor world of chaos, in the beginning of the present order of the universe, was fashioned in the mathematical manner assigned. The material was rolled into a spheroid, and sent whirling on its axis around the common centre of the solar system.

Gravity could act; but the laws of life could not people the fiery globe. They existed, ready to fashion living beings; but they were prevented from acting by the form matter then assumed. Pure motion must labor longer. The fiery vapor must condense by radiation. A crust must form over the liquid globe. Water must be created; and then only in that universal menstruum could the organic laws act their part, and create an individualized life from the life of the globe, evoked by gravitation. The material on which it acted was rude, and hence an imperfect result must be the consequence. A perfect being could not at once spring into maturity on the vacant globe, like Minerva, in full possession of all powers, from the brain of Jove. Such might be expected did special design and miracle govern the world, and did not law hold regency in the domain of Nature.

A few cells floating in the dark ocean, stretching among black ledges and craggy pinnacles, proclaimed the advent of life.

Slow and painful was its progress through Eons of ages, decadecillions of infinite epochs, every step rising higher as the physical world became refined and adapted to its sustenance. Higher and higher it advanced, until the earth became prepared to bring forth and sustain man; and he came. The tree of life put forth its flower in the exceeding perfection of its structure. Equally painful has been his progress from brute to savagery, from savagery to civilization.

Now, I hold this as self-evident—that if a cause external and superior to matter had been exerted, all this history would be swept away. We should have the old mythological tale of a Garden of Eden, and perfect man at once placed in a perfect world; and to harmonize that state with present stern and deplorable reality, we must mythologize further, and invent his fall, and, as balm to this mortal wound, his consequent redemption. That such is not the fact, proves incontestably that no such cause has ever been exerted; and we are forced to adopt the doctrine of immutable law, making changeable matter obedient to its sway. Hence the theory of slow but eternal progress of life, of the globe, of matter itself, which, underlying every thing else, forces all onward.

CHAPTER XIV.

PHILOSOPHY OF CHANGE AND DEATH, CONCLUDED.

A Clairvoyant Revelation. — A Death-bed Scene. — Parting of Spirit and Body. — Spiritual Experience. — What they say of the Middle Passage. — Revelation of an Atheist. — Of a Spiritualist. — Robert Owens. — The Arcana of Death disclosed.

[ONE calm and beautiful winter evening I became entranced. A voice whispered to me, "I will show you now the philosophy of death." I seemed to leave the body. I existed outside of, and independent of, the physical form; yet I observed a connecting line uniting my spirit to its shrine. The spirit took my hand, and we passed from my room into the air. The stars shone beautifully from the icy arch, and the moon flooded the landscape with a deluge of silver light. Silently in slumber, wrapped in its gray mantle, lay the weary earth. We seemed the only living beings of the shadowy landscape. On we passed with the swift wings of thought until we came to a palatial dwelling. A light feebly shone from a single window, speaking of disease even in that sumptuous residence. No bell announced our arrival, no knocker rang through the hall. The window furnished an open way, and unannounced we entered.

On a couch of softest eider lay a beautiful child, just blushing into womanhood. Disease had wasted the physical form until her spirit stood so far across the threshold of the spirit world as to cast over the dying clay the radiance of heaven.

The rose had vanished, but her eyes spoke volumes of angelic love, for they already saw the bright spirits around her. They met the fond expression of a grandmother and a sister, ready to receive her in their extended arms.

At her side her mother bent beneath the intolerable weight of grief, and at the foot of the couch stood her stern father, his pride subdued by wretchedness. It was heart-rending to witness the scene. For death is a grim monster, whose jaws receive our fondest loves, and hide them from our view forever; and unless we are imbued with the spiritual philosophy, dark indeed is the gloom which hangs like an impenetrable pall over the grave.

A holy radiance stole over the face of the dying girl. She extended her hand as if to grasp another's.

"How beautiful!" broke from her pale lips. "I come;" and she went to those who awaited her.

The wasted form still reclined on the sumptuous couch, but the light of the spirit was gone. Dark and dreary was the scene in that apartment.

But what was the process by which the spirit was freed from its earthly body, and ushered into the next plane of its existence? Very simple and very beautiful. It was a higher degree of clairvoyance. Slowly the spiritual form withdrew from the extremities and concentrated in the brain. As it did so, a halo arose from the crown of the head, which gradually increased. Soon it became clear and distinct, and I observed that it was the exact resemblance of the form it had left. Higher and higher it arose, until the beautiful spirit stood before us, and the dead body reclined below. A slight cord connected the two, which, gradually diminishing, became in a few minutes absorbed, and the spirit had forever quitted its earthly temple. New faculties were bestowed, new and dazzling sensations experienced, and the grand spheres of spirit life darkened the mansions of earthly pride.

Thus I investigated this awful subject, which in the clouded minds of all exerts such fear and horror. Death has long been looked upon as a dreadful gulf, which divides the mortal life, perhaps, from oblivion, the vale of tears and sorrow, where man's noble faculties perish in the darkness of eternity. Those who profess unflinching faith in Christianity are disturbed by fears and uncompromising doubts, and see little hope for an existence beyond the "narrow house." No tidings are borne across the dark river. The promised land is a "bourn from which no traveller returns" to tell its tales of joys or sorrows. A heavy veil of mist hangs over the rudimental sphere in regard to the great change all must meet when the body becomes worn and wasted, and the soul trembles on the brink of the awful gulf, which, it is taught, once passed, could never be repassed.

With these dark clouds encompassing the departing spirit, death is feared as the fell destroyer of the race, and under these impressions the safe and easy journey is a *real* gulf of anguish.

After my clairvoyant view of the sad yet joyful scene with which I began this chapter, I received communications from several spirits descriptive of their sensations at the approach of death. Some of these may be interesting, as they illustrate the grand philosophy we are striving to set forth.]

While a resident of earth, I was indoctrinated in the

religious absurdities which prevail in that sphere. I was taught to believe in a personal God and devil, one having supreme control over heaven and the other over hell; and still more absurd the mission of Christ. He came not, as I supposed, to forgive sins, but as a Reformer, to point the way.

It was after a life spent in the pursuit of worthless objects that I lay on the couch of death, and my thoughts awakened to unusual activity. I thought of all the past scenes of my life, and the frightful gulf I was soon to pass. As I thus reclined in gloomy thought, not a single star presented its beacon light to give me hope. Kind and regretful friends stood weeping at my bedside. O, how I desired to speak one word, and tell them not to wring my soul with anguish by their tears. But I could not utter my request. Not a word passed my frozen lips.

I had no treasure over the unfathomed gulf. Though my wife had gone before me to heaven, yet I did not suppose I should recognize her. Consequently I had no treasure in the spirit-land. From the dark picture of my sorrowing friends I turned to one still darker. The dreadful gulf I was fast approaching, presented an appalling aspect, which it seemed impossible to endure.

* * * * * *

A deep sleep enshrouded my faculties. During its continuance I neither saw nor heard any thing which passed around me. This I had since learned was the sleep of death which I had so much feared — the gulf which had caused me so much anguish. After slumbering an indefinite period, I awoke into life in another sphere. A holy, sacred light pervaded all objects, and a halo-like glory emanated from every thing I saw. The first object I saw was she whom I once called my wife. She spoke to me in tones of love. My astonishment was boundless, my joy equally great, for it seemed that she had returned from a long absence to greet me with her love.

I looked around me. Below, and seemingly a part of myself, lay a form of earth, cold, stiff, and motionless. Around it stood friends weeping for a departed brother. That form I recognized as my own. The sorrowing I recognized as my friends'. And yet, although separated from my body, I was myself. While I was reflecting why this was as it appeared, my guardian spirit whispered, "You have crossed the dark chasm of your imaginary terrors, and are in the land of spirits."

I answered that heaven could not be on earth; but she replied that "heaven was where there was a happy mind." I asked her again and again where I was, to receive the same response.

Still my friends sorrowed at my bedside; and, while contemplating the strangeness of the scene, I first became aware that I was invisible to them. I could see them, but they could not recognize my presence.

My guardian said that brighter scenes of beauty awaited me, and beckoned me to follow. We seemed to tread upon the airy flood, and not to be subject to the laws of gravitation. We floated out on the polar current into the unlimited ocean of space.

* * * * * *

My thoughts were aroused into extreme activity. I looked around, and first began to realize and enjoy the second state of existence. The reality of that state I enjoyed; gazed on its radiant sun which filled the atmosphere with its beams, and was received as a brother by a society of spirits who were harmonious in their desires.

Here I enjoyed the love of kindred spirits. I rapidly arose from my former position, and all my errors were gradually exchanged for truth. I have often considered it very strange that thinking beings, enjoying the light and perfections of nature, should fall into such egregious errors as I did while on earth. But when I behold, every hour, minds receiving superior light falling into more absurd errors, I cease to wonder at my former ignorance. Here is no contention, but peace and harmony — not hell, but happiness. My punishment for my errors was the shame I experienced at my former delusions, which I knew all spirits discovered when they beheld me. But this soon passed away, and I was happy.

The second communication which I introduce is of an entirely different character.

I was an atheist. How I came to assume that position may seem strange; but the same reasons have convinced the understanding of the majority of mankind.

I saw the Bible as a book claiming my belief for the truth it contained, and no further. But those who set themselves over me as religious teachers, said I must believe it all, *verbatim*, as written, or not at all. My reason declared that many of its passages, scattered here and there through every chapter, did not agree with nature; and to believe contrary to reason and nature I could not compel my faculties. Nor would I declare to the world I believed when I did not, and being by force of public opinion obliged to condemn all or believe all, I chose the former, and supported my position by nature, as I understood her revelations. The position which I had taken caused to fall on me the determined hatred and scorn of the sects who pretended to believe the book they had caused me to condemn, and who were the professed followers of the Reformer sent from heaven — as they believed — to teach mankind love and peace.

This confirmed the doubts I had previously entertained, for instead of coming to me in the spirit of the one they professed to worship, they shunned and avoided me.

After a life passed in research into the concealed laws of nature, and the laws which govern the external world, I reclined on the couch of death. No dreadful gulf was to be passed, no frightful scene to be enacted. My mind was peaceful and quiet, for I had done my duty. I felt the calm resulting from an upright life. Soon I was to pass from earthly scenes forever, becoming as though I had never been. Like the animal whose existence terminates at the same point where it commences, so I supposed it would be with me; for we might as well expect the hum of the bee when the insect had passed, as life after the body was dead. This was my philosophy, and from my material standpoint I could see none more reasonable.

I felt the dreamy sleep approach. My senses were entranced; my speech was gone; I knew I was dying. I slept a dreamy slumber.

* * * * * *

After an indefinite period passed in oblivion, I

awoke to life. A divine glow pervaded all objects; my thoughts expanded; oblivion was passed; my spirit ransomed, yet I knew it not. My belief was unshaken, and I still awaited the approach of death. I looked below me on a rigid corpse, which I knew was myself; yet I was an identity. My mind was confused and bewildered. It seemed as if I was dreaming; yet the strange scene and its reality convinced my understanding. I disbelieved what I before believed, and believed that which I had always regarded as folly. All my theories I knew to be false, for there was a reality beyond the grave.

It was years, however, before I entirely overcame the idea that possibly I might be dreaming. After I had mingled with the millions of earth's departed, and beheld the grandeur of this sphere, the vague idea of my living a dream-life would unavoidably steal over me.

But did you not receive punishment? No more torment, no severer punishment, than that of remaining in the belief that I dreamed, and the unsatisfying state of mind I possessed on earth. We know no hell but that which is in the mind — no devil but that which every man becomes unto himself.

A third spirit in a more ecstatic manner speaks of the change.

As I write this I cannot but exclaim, How wise in wisdom, how benevolent in benevolence, how lovely in loveliness, how incomprehensible in incomprehensibility, the great whole is fashioned! I dwell in a boundless ocean, in which suns and worlds are continents and islands. My soul is unlimited in its extent. My spirit seems of gigantic proportions, and capable of grasping nature's most abstruse laws. I am free from all errors, and truth is the object for which I continually strive. My death was the same as will cause many to leave their rudimental homes; they become so far developed that the rudimental form will not serve the spirit. The cerebrum and cerebellum will not retain the spirit after it has attained a certain degree of refinement, and as mind advances towards the perfection of the sphere in which it is retained, it bursts through the conditions of that sphere and ascends to a higher. My mind had reached such a condition. I desired a higher state, and my desire was gratified.

The night was beautiful — too enchanting for one to depart from earth. The fair orbed moon threw a halo of light on the night-side of earth. The entire scene would have chained one whose faith in the reality of the future state was less strong. I longed to bid adieu to the scenes of earth.

Closing with the day the affairs of the sphere, I retired to rest. The deep study of that day had so expanded my mind, that my brain would contain the spirit no longer. Every little heart or centre, instead of throwing its vitalizing fluid to the distant parts of my body, drew it into itself. It seemed as though my brain would burst; yet the sensation was not painful. I knew that I was dying. I was deeply impressed with the fact, yet rejoiced that I soon would leave this world for a higher sphere; not leave earth forever, but no longer mingle in its jostling throng. I was certain of a hereafter, and rejoiced at its approach. The scattered light reflected through the medium of clairvoyance had so informed me that I feared death no longer.

I felt my spirit arise from my form, issuing from my head, until only a slight line connected me with earth. After perhaps an hour, this divided; then I was free. My home was no longer on earth, but in the spirit-sphere. But how can I describe the sensations produced by nature on spiritual vision? I can only compare them to those of the blind man when suddenly his eyes are opened, and the glorious beams of the sun burst on his senses, so intense were my impressions.

I thank fortune that left me alone to die. No weeping friend at my bedside to call me back and tie me to earth. The sacred silence was unbroken, and I was free to obey my impulses.

By death I have realized the truth of a future state, before faintly shadowed. My present occupation and enjoyment is visiting the various worlds of space, and investigating the laws by which they are governed. While on earth I greatly desired to travel; I now can gratify myself, and not only visit earth, but also all the innumerable worlds which are concealed in the depths of space.

I also realize the beauties of the future life — that nature was created for man, and man for nature, and should perfectly harmonize; that there are as many natures as individuals, each harmonizing with its appropriate objects; that all are parts of one great whole, bounded by immensity.

My brief history is terminated. By it you will know that if mind understands its future destiny, death will not be painful, but a pleasant journey from earth to the spirit land, or an earthly slumber and spiritual awaking.

As further illustrative of this interesting subject, I

will introduce a short article from my friend Robert Owen, given soon after his death. [The reader must bear in mind that each spirit is held responsible for his own sentiments.]

"Though we never met in the body, I was strongly attracted to you, and received with pleasure the letters you wrote in answer to mine. Most sincerely do I thank you for that heartful communication which I received, previous to my death, from my most respected and esteemed friend, Dr. Hare,* of the spirit world. You know not the comfort and pleasure I received from it. True to his promise, he was the first to welcome me to my new home. His youth was renewed, and joy, pleasure, and goodness beamed from his countenance as he benignantly gave me his hand, and raised me from the ruins of my mortality.

"How blessed is spiritualism! Had I died in my infidelity, most dark, painful, and cruel would have been my transition; as it was, it was sweet, pleasant, and joyful. I had lived more than my allotted time on earth — so long that my body had nearly perished atom by atom. But I accomplished a great, noble work, such as no other individual could or would have ever accomplished. Peculiar circumstances and influences developed in my mind ideas none other ever possessed; and for half a century I labored to build a social fabric, which, in goodness, purity, spirituality, brotherly affection, and practical benefits, should shame the old, time-worn, and obsolete systems of the world founded in sin, error, and corruption. For fifty years I labored, unfalteringly, happily at this task. You

* Alluding to a communication from him.

may say I accomplished little. So I sometimes thought; but I now see that a great good has and will grow out of my efforts. I appealed to the wrong source. I petitioned and memorialized the rotten, tyrannical governments of the world, instead of man himself. I scorn the idea now.

"In this sphere those three great curses which I combated while on earth are removed. I found a sufficient number to afford me sympathizing association such as I never dreamed of. Those three plagues of the world are its superstitions, by which it tramples on the weak martyr, and crucifies its saviours, to glorify them in succeeding ages, like an ignoble fool; legalized marriage, consigned to the mummery of a priesthood, from which arise all the prostitutions of the world and the degradation of the female sex; and private property, the distinction of mine and thine, from which arise all the robberies, frauds, falsehoods, and crimes of the world. Against these I have ever waged war, and ever shall until they are overthrown. I find this is my heaven. Surrounded by a group of kindred minds, we all, as one, strive to perfect a social system which we shall impress on the impressible of earth's inhabitants, and endeavor to actualize in the world life. Let kindred spirits be drawn together in harmonious groups; let them be surrounded by proper conditions, and crime, error, and folly would rapidly give place to goodness, love, virtue, and general peace. War would perish, kings and rulers cease to be, love and joy reign over the delighted people.

"Of this I shall write in detail another time, and in a more fitting place. Other spirits have given my experience at death when they wrote their own. No one, however, can sufficiently appreciate the value of the boon conferred by spiritualism. Belief is every thing while crossing the 'dark river.' One spirit voice converts the darkness into light, in which the forms of loved and cherished ones appear. ROBERT OWEN."

Such are the arcana of death. It is not a fearful but a pleasant change; from the chrysalis state the spirit bursts into full maturity. The earth is the infant school where the spirit prepares for eternity, and the infant sustains the similar relations to the man that the man sustains to the spirit.

But the transition is not always accompanied with such surroundings as here described. Often men are buried beneath avalanches of rocks and ice, or miles of ocean; in deep wells, or bottomless fissures. How can the spirit in such cases' depart? Does physical matter wall it in as it does the body? This cannot be. As a crystalline body transmits light, so matter transmits spirit. As one is transparent to light, the other is transparent to spirit, and offers no obstructions to its passage. The solid rock, or the wall of a room, does not offer any resistance to the passage of a spirit more than the thin and yielding air. The relation between matter and spirit, or spirit and matter, is not the same as exists between spirit and spirit, or matter and other matter. Spirit holds the same relations to spiritual things as man holds to physical, but gross matter is to them as a nonentity so far as it offers obstruction to their progression.

With these illustrations the philosophy of the great change becomes comprehensive. One thing is clearly ascertained — death does not change the mind, but only the body, from which the mind is withdrawn.

In whatever position the spirit was when death came, there will it be in the future, until it progresses from it, how far it is advanced in love and wisdom when it sinks into the clairvoyant sleep of death, there will it awake, and at the precise point where the mortal fell asleep, there will the immortal commence a new life, with all its former acquisitions, and no more.

Death, thus divested of the terrors with which mythology has invested it, is but a pleasant journey from one clime to another, painless and sweet; a peaceful sleep, silent and profound; an awaking of the spirit in the spirit land. When the muscles contract it is not with pain, but by the changing electric equilibrium, induced by the departing spirit.

Man, when matured by age, dies, as the ripe fruit drops from its stem. Death begins when the body commences to deteriorate, and when sensation ceases its task is ended.

Such is the gateway to eternal life — the sleep of the body — the heaven of the spirit.

CHAPTER XV.

SPIRIT, ITS ORIGIN, FACULTIES, AND POWER.

What is Spirit? — What is its Origin? — Value of Metaphysical and Theological Knowledge. — True Method of Research. — Microscopic and Clairvoyant Revelations. — Circumstances of Birth of the new Being. — Office of the Sperm and Germ Cells. — Their Union, Results of. — Further Growth of the Germ. — The Dual Structure of Man. — Intuition as a Guide. — An Anecdote from St. Augustine. — Plutarch's Opinion. — The Problem of Man's Immortality a vexed Question. — The Doctrines of Cause and Effect introduced into the Realm of Spirit. — Proof that the Spirit retains its Form and Senses. — Clairvoyant Testimony. — Our own Evidence. — The Spirit Body; its Relation to the Physical; its Fœtal Growth; Period of Individualization. — How far must Man be developed to become immortal? — Beasts mortal, and why. — The Line of Demarcation between Mortal and Immortal Beings. — Necessary Conditions of Immortality.

What is spirit? Is it an imponderable, intangible nothing, capable of thinking, of feeling, of condition? Is it a "detached intellect"? Little is known by earth of its existence. The boor and the sage stand here on nearly the same ground; for it is as impossible for man to *comprehend* his future existence as the caterpillar to understand its butterfly state. Month after month it dwells on the gnarled branch, gnawing the acrid leaves, no change passing over it. Its sphere is humble and circumscribed. It rarely leaves the branch on which it sprang into existence. There it weaves a winding sheet around itself, and passes into the unconsciousness of death. It awakes, bursts its fetters, and is resurrected a beautiful winged gem, basking in the warmth and light — a reveller among the flowers. Splendid type of the soul, when it casts off its fetters of clay, and feels the invigorating breath of the spirit-sphere. In this inquiry we discard metaphysics, and all previous systems of psychology. We enter a new and unbeaten path. We pass by the old landmarks, Columbus-like, leaving the known shore behind, and guided only by the compass of Reason, steer boldly out into the ocean of unknown truth. Have a care, friend, how you receive our words. Let us not lead you astray from your reason.

The origin of the spirit has perplexed the thinkers of all ages. The creation of the body in the maternal matrix could be observed in all its progressive stages; but the spirit, being invisible, could not be thus studied. The question has been ever asked, Whence come the immortal principles?

The theological answer eluded the dilemma by characteristic shrewdness. God creates the spirit in heaven; and when opportunity offers, it enters the germ, and is clothed with an earthly garb. This solution is worthy the dark ages, but in no wise can be at present admitted; for we know law, not miracle, rules Nature. According to this conjecture, the birth of every new being requires a direct miracle; and we may well ask, Why do we lose the consciousness of our previous states? Is our earth life a dream life? Shall we never know an actual?

Science laughs at such idle schemes, and the word-wrangling of metaphysicians, while it lays its sure foundation by facts. What we know we must know positively; and the metaphysical and theological knowledge of the schools is worse than folly — worse than the absence of all knowledge.

To study this important subject, we must commence at the same point where we would to study the origin of the body. If the spirit exists, it is supported by law; then it originated by law, and its intimate relations to the body presuppose a common origin. The microscope looks far into the mystery; but clairvoyance sees further, and with clearer eyes. If we employ either method, we shall find that the first intimation of the new being is the birth of two cells.

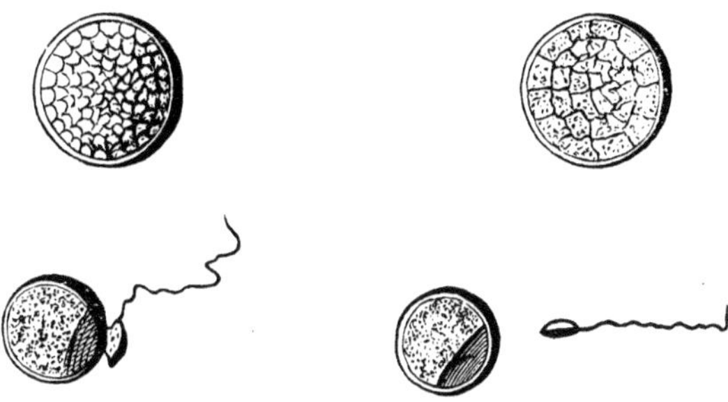

Union of germ and sperm cells. An advanced stage.
a. Sperm cell. *b.* Germ cell. *d.* Their further advancement.

In Fig. 1 we represent these two cells. *a* is the sperm cell, originating in the male parent, and concentrating his spiritual and mental being in a focus, as a burning glass concentrates rays of heat or light. It is to appearance a simple vesicle, with a vibrating celia or fibre, which seemingly is endowed with a slight degree of life, or has a partial progressive motion. The contents of this vesicle are extremely luminous, and we may suppose it has the property of condensing the psychological influence of the male; although from its extreme minuteness, credulity even might pronounce the idea preposterous. Several thousand would not form a point larger than a mote dancing in the sunshine. Yet we are obliged to consider it as the vehicle of transmission; for in no other manner can the offspring partake of the character of its sire.

b represents the germ cell, — the product of the female, — much larger than the sperm cell, and without a celia. It at first is composed of one cell; but this divides, and these divide again, and so on *ad infinitum*. In the figure, *c* shows the first stage of this process; *d* represents its farther advancement. In this respect it exactly copies the process of growth of the lower animals. Its growth has no relation to the development of organs, but is simple extension in all directions by the multiplication of cells. If this process is not interrupted by the contact of a sperm cell, it is continued for a short time, when the inherent energy of the primary vesicle is exhausted, and it is thrown off.

But if a sperm cell is brought near the germ cell in its early state, they rush together as opposite poles of two magnets. The formative principle combined in the sperm cell is exerted on the germ. The plastic material yields to the hand of its master. The material exists in the germ cell; but the formative principle is absent. There is nothing to direct the infinite multiplication of cells into a determinate channel. Like the gigantic puff-ball which springs from the rank soil, and in a single night multiplies the cells which compose it by millions, to perish with the rising sun, so the germ grows and perishes. The channel of its growth is proscribed by the character of the sperm cell, that imperceptible vesicle.

We need not introduce facts to support this proposition, for all know how indelibly the character of the sire is stamped on the physical and psychical nature of the offspring.

Fig. 2 shows the attachment of the sperm to the germ cell, and in Fig. 3 we represent a considerable advanced stage of growth. Each cell in the conglomerate mass represents the rudiments of some organ, which will bud forth by its excessive multiplication.

The growth of the fœtus from this point to its birth is minutely described in the preceding volume, and hence we shall only glance at the growth of spirit.

If the parents have immortal spirits as well as mortal bodies, and if their corporeal frames support the corporeal being of the fœtus, then their spiritual natures must in an equal measure support the spirit of the fœtus, and the growth of its spirit and body be similar, both receiving nourishment from the mother. This chain of argument cannot be broken.

After birth, the five senses bring the external world to an intense focus in its spirit.

This philosophy makes man a dual structure, an internal and external being united. The old division of body, soul, and spirit we discard. Soul is a meaningless term, representing nothing unless used in the place of body or spirit—a use leading to confusion and erroneous ideas. If employed to represent the blood, or life, as it has been by many authors, it conveys a wholly erroneous conception, for the blood has no more right to the term life than the bone or muscle. Hence we have substituted the natural order for the artificial division which was first introduced and has ever been maintained to make a show of knowledge where profound ignorance reigned.

The human mind, guided by intuition and observation, has always followed more closely the path of truth than when guided by the lurid light of its pretended revelations. The ancient Greek philosophers entertained a more correct idea of the future life than the Hebrew, or even than is contained in the New Testament. The early fathers, imbibing the Greek philosophy, taught a far more spiritual doctrine than is contained in the text. We cannot refrain from inserting a short quotation from St. Augustine.

" Our brother Sennardius, well known to us all as an eminent physician, and whom we especially love, who is now at Carthage, after having distinguished himself at Rome, and with whose active piety and benevolence you are well acquainted, could not, nevertheless, as he related to us, bring himself to believe in life after death. One night there appeared to him in a dream a radiant youth of noble aspect, who bade him follow him; and, as Sennardius obeyed, they came to a city where on the right he heard a chorus of most heavenly voices. As he desired to know whence this heavenly harmony proceeded, the youth told him that what he heard were songs of the blessed; whereupon he awoke, and thought no more of his dream than people usually do.

" On another night the youth appears to him again, and asks him if he knows him; and Sennardius told him all the particulars of his former dream, which he well remembered. 'Then,' said the youth, 'was it while sleeping or waking you saw these things?' 'I was sleeping,' answered Sennardius. 'You are right,' replied the youth; 'it was in your sleep that you saw these things; and know, O Sennardius, that what you see now is also in your sleep. But if this be so, tell me where, then, is your body?' 'In my bed chamber,' answered Sennardius. 'But know you not,' continued the youth, 'that your eyes, which form a part of your body, are closed and inactive?' 'I know it,' answered he. 'Then,' said the youth, 'with what eyes see you these things?' And Sennardius could not answer him; and, as he hesitated, the youth spoke again, and explained the motives of his question. 'As the eyes of your body,' said he, 'which lies now in bed, and sleeps, are inactive and useless, and yet you have eyes wherewith you see me and those things I have shown you, so after death, when these bodily organs fail you, you will have a vital power whereby you will live, and a sensitive faculty whereby you will perceive. Doubt, therefore, no longer that there is a life after death.'" *

A great truth is contained in this episode. Man is a dual structure — a spirit and a body blended into a unit; the body related to the external world by the senses, the spirit taking cognizance of the spiritual world through its spiritual perceptions. The spirit is the companion of the body, and as long as the two are united it perceives the relation of the external world through and by aid of the corporeal senses. So much is the spirit concealed by the corporeal body, so intimately are they blended, that it is with difficulty its existence is perceived.

Plutarch well observes, in the strict spirit of induc-

* Epistles, 159, Antwerp edition.

tive philosophy, that if demons and protecting spirits are disembodied souls, we ought not to doubt that those spirits, when inhabiting the body, possessed the same faculties they now enjoy, since we have no reason to suppose that any new faculties are conferred at the period of dissolution; for such faculties must be considered *inherent*, though obscured or latent. The sun does not first shine when it breaks from behind a cloud; so the spirit, when it first throws aside the body, does not acquire the faculties which are supposed to characterize it, but they are only freed from the obscuration of the mortal state as the sun from the fetters of the cloud.

The problem of man's immortality has been vexed from immemorial time, yet the theologian and metaphysician, after all their gigantic efforts, have accomplished nothing by way of demonstration. They have never met the question fairly, and scanned it by the light of natural law. Forced to admit certainty into the domain of the physical world, — a term by which we mean what they understand as the world of matter, — they have ever regarded with holy horror the introduction of cause and effect into the realm of spirit. On the threshold of this realm the inductive philosophy, that magnificent system which traces effects to their causes, which discerns a cause beneath every effect, has been dismissed as a profane and erring guide, and in its place a will-o'-the-wisp led them through the reeking miasm of metaphysical controversy and along the slippery paths intersecting the night-enveloped swamp lands of bigoted and insane theological disputation.

One fact in clairvoyance — one manifestation of

spirit presence — outweighs all the logical argumentations the world has ever heard. How far these support immortality a preceding chapter has shown.

We said that if spirit existed it must have form. It must retain, whatever others it may acquire, the five senses. It must be organized. Let us investigate this proposition. The clairvoyant has entered the deepest trance. His body lies oblivious, as near the portals of death as it is possible and not enter. All avenues to the senses are closed; the blood flows slowly and turgidly along its channels; the nerves have lost their irritability; and the brain cannot feel. The blinding lightnings affect not the eye; the crash of thunders are not heard by the ear. Limb after limb can be severed unfelt. Such is the state of the body. What is that of the spirit which has thus temporarily deserted it?

Not unconscious, not senseless, not inactive, but like a freed eagle it soars in the light of a new existence. The channels through which it obtained a knowledge of the world of matter are closed, it is true; but it has no necessity for them now, for spiritual light acts on the spirit eye, waves in the spirit atmosphere vibrate on the spirit ear, and feeling becomes a refined consciousness, which is more delicate and exquisite than it possessed in the body by all conception. It sees, it hears, it feels, while the body can be burned to ashes without pain, or even automatic irritability.

With such facts before you, how avoid the conclusion that spirits are organized, and that they have all the senses as perfect, nay, more perfect than man?

But why this argument? Do we not exist? and

are not all our faculties and senses preserved? To us these pages are self-evident truths; and it is like man's endeavoring to prove that he sees, hears, or feels. But until our evidence is considered positive, argument and induction must be employed; and though we guide the pen which indites these sentences, we must submit to the humiliating task of proving that we have a substantial existence, and are not phantasies of a disordered brain.

The spiritual body is matured with the physical form which it pervades; and when the latter is cast off we find it existing independently. Admitting this view, the question arises, Where is the line of demarcation between a mortal and immortal being drawn? The beast, that surrenders up its existence to the bosom of universally diffused spirit, and man, who conquers oblivion, — in what consists the difference?

These questions lead us still farther to the investigation of: What constitutes an immortal being? Why is the physical organization mortal? Why is death an essential condition of life? Organic forms decay, because in them the waste exceeds the renovation. The vital system wears faster than it is repaired, and of necessity vitality is suspended. If the vital organization, and all conditions surrounding it, were perfect, there would be harmony between all vital processes, and death never result.

In the chaotic ocean of the beginning, Nature strove to bring all causes and effects into harmony, and in proportion as this was accomplished the earth became beautiful in its order of arrangements. Such is the voice from its historic pages — the rock-volume we have sought to translate.

Nature is striving for harmony, and she approaches this state in the perfect man, with his great yearning for future life. Immortality cannot be bestowed on him through the agency of physical elements. They are too gross to form a perfect unity, and a unity harmoniously preserved is the prime necessity of immortal life. In the ethereal realm of the spiritual elements such a harmony can be produced, and organizations formed in which renovation and decay exactly balance each other.

Such is the ultimate effect of creation. Man is the crowning glory of Nature, and the crowning glory of man is his immortality. Destroy this, and all the labor of Nature is abortive. The greatest mistake possible to commit has been committed, and the grand scheme of creative energy is a total failure.

Animals as well as man have spirits, but they are not immortal, for even in such ultimated elements harmony cannot be maintained after the death of the body. To illustrate this idea, an arch may be built never so perfectly, but if the keystone is not put in place the whole will fall in ruins so soon as the staging is removed; but lay that single stone in place, and the whole stands firm as a rock. So with the spirit of the animal: it is an imperfect arch, which so soon as the body which supports it is removed falls. But the spirit of man is a perfect arch, standing firm after the removal of the body.

But as the animal merges through intermediate forms into man, and the infant knows less than the perfect animal, the line of demarcation before alluded to apparently is drawn with difficulty. Not so, however. A certain degree of refinement is absolutely essential, below which is nonentity, above which is immortality — not sharply drawn, however. A spirit is not necessarily immortal, but can be gradually extinguished, as a lamp, burning for an indefinite time and then slowly going out. Such is the condition of the lowest races of mankind. Their spirits exist after death, but in them there is no progress, no desire for the immortal state, and slowly, atom by atom, they are absorbed into the bosom of the universal spirit essence, as the spirit of the animal is immediately after death.

It may be asked, At what age does man become immortal? No certain time can be given, as no sharp line exists, but the time varies according to the infant's development.

The idiot — is he immortal? This is a very inaccurate question, for the answer depends on circumstances, as degree of idiocy, causes, &c. If destitute of a ray of intellect, a voiceless, thoughtless idiot, the inference is not cheering; for if existence be preserved after death it will probably be absorbed in a short time, as the Hindoo would say, into the infinite bosom of Brahm.

The inference from the foregoing is, that a spirit can become mortal, and lose its individuality. This is true of spirits considerably advanced. It can, by a course of debauchery, gluttony, lust, and crime, annul its charter to immortal life, and gradually fade into oblivion.

But our pen turns from this cheerless picture of wretchedness to brighter scenes. We are pained to write these sentences, which sound like the doom of guilt pronounced at the seat of justice against the capital offender; but we are pledged to present the ultimate workings of Nature's government, joyful or cheerless, for we know the end is good. The spiritual elements, which result from such dissolutions after this cycle, return to other growing forms, are embodied in other spirits, and perhaps this time are individualized in noblest immortals.

While some spirits go out of existence, as the ill-trimmed taper, the destiny of others is inconceivably grand. Passing onward from sphere to sphere, becoming more and more refined, more elevated in thought, perfected in wisdom and understanding, they enter celestial and supercelestial spheres, becoming far wiser than finite comprehension of God. Where the end will be we know not; but this we know—that suns and systems will fade, the great universe itself will pass like the illusions of a dream. Still will the spirit, less and less fettered, vault upward, rejoicing in increasing strength and wisdom.

CHAPTER XVI.

A CLAIRVOYANT'S VIEW OF THE SPIRIT SPHERE.

Description of the Sensations when entering the Clairvoyant State. — Why not terminated by Death. — Floating on a Magnetic River. — View of the Sphere. — Scenery described. — The Mansion. — Occupation of its Inmates. — Return.

How can I describe the sensations I experienced when I first sank into the superior clairvoyant state? I cannot. Words are employed to convey known ideas, but the ideas there awakened have no words, and must remain unexpressed.

I was communing on a deep topic with my spirit-friends, through my impressibility, and writing the words as fast as they were given me, when I perceived that the sweet sensation, which fell like a gauzy veil over my nervous system, was slowly deepening. Before I was aware, earth objects were excluded from my senses. I saw not, heard not, felt not, and my tongue refused to speak. It was like dying. The blood seemed to flow in from the extremities, and concentrate in the heart and brain, and the former organ soon partook of the paralysis, beating slower and slower, until its pulsations were imperceptible.

My mind grew active. It threw off the restraint of the body; a thousand rainbows came and went in rapid change of intermingling haloes, with all the beauty, diversity, and rapidity of the kaleidoscope. I felt myself arising from my body — felt that I was free; at least felt not the weight of its physical fetters.

Then my mind was quickened. Thoughts grand and inexpressible came like pulsating waves from every side, and it seemed that I was *en rapport* with the combined intelligence of the angel sphere. Not till then was I aware that by losing my physical senses I had acquired spiritual perceptions infinitely more acute. The scene which before spread around me dull and monotonous brightened with spiritual radiance. The colors were vivid and gorgeous, and an ethereality involved all in a dreamy maze.

While transported with exulting rapture at the beauty of the change, I became conscious that a person was by my side. I turned to look at him, and recognized a guardian spirit. With a beneficent smile he took my hand, saying, "Son, I am thankful for this opportunity to show you the reality of our existence."

I could not answer, but he read my grateful thoughts. For a long time I could not take my eyes away. His lofty brow was shaded by snowy locks of glossy hair, and his white beard waved on his bosom. His keen blue eye spoke of long centuries of deep investigation into the mysterious labyrinths of nature, yet overflowed with ineffable kindness. His robe was silver white, and fell around him in delicate drapery. Tall, noble, and spiritual, he stood before me, with his great heart speaking of universal love and benevolence.

"Observe this small cord of spiritual matter, which passes from your head to your physical brain," said he.

I looked, and saw what I had before failed to notice — a small silvery line of particles flowing to and fro.

"That," continued he, "is all that connects you

with earth. If it became broken, death would immediately ensue. You have suffered all the pangs of death: you have taken the successive steps; and sever this, and you would take the last."

"But," asked I, "is not death more painful?"

"No," he answered; "death is never painful. The disorganization which produces it may be severe; but when death begins its work, the organization is lulled to rest with an opiate overpowering disease itself. The contortions of the muscles, which frequently attend it, are the result of the disturbed electrical conditions caused by the withdrawal of the spirit, and are not attended with pain. You have suffered all that is ever suffered in dying. The rupturing of this thread would not be painful, and then your body would moulder back to earth, and your spirit remain in this sphere forever."

"Why is it not broken? Why is not the process carried farther, until death terminates the clairvoyant process?"

"Because there is harmony between your spirit and body. They are adapted for each other, and between them exist the strongest attractions. True, your spirit has left your body, but it has not withdrawn the vital magnetism necessary to keep the physical mechanism in motion, and preserve it unimpaired against your return. A close sympathy is preserved by this cord, and nothing but a disaster occurring to your body can break it."

"What a beautiful work!" I exclaimed.

"Let us go," said he, taking my hand.

We arose from the floor of the room, through the ceiling and roof, and soon were far above the earth

I thus became aware that physical matter offers no resistance to the passage of spirit organisms. They can pass through the walls of a house, through a solid rock, or into the earth, as easily as through the atmosphere. They are borne up and supported on all sides by a spiritual ether, which pervades all space, and all bodies, and wherever that enters they can go. Their organisms being lighter than this ether, their gravity is annulled, and they fly in any direction by the force of their wills rendering them positive to their destination.

As we rapidly passed to the north, I felt a strong impulse bearing me onward.

"What does this mean?" I inquired.

"It is the flow of a vast river of magnetism, which passes from earth to the spirit sphere," he replied. "It passes, you see, parallel with the earth's surface, until it reaches the polar opening, when it arises and becomes diffused in the spiritual atmosphere."

Above us spread the spirit sphere plainly discernible, as a broad belt extending each side of the equator, sixty degrees, and hence covering the whole southern heavens.

"Why does not this thick belt conceal the sun and stars, and thus manifest its presence?"

Because it is composed of spiritual matter, which holds a different relation to light than that held by physical atoms. Light is composed of numberless elements, and while this zone intercepts the spiritual portion, which lights its surface, it freely transmits that portion which is light to earth. Our spiritual eyes are so organized that the earth light is darkness to them, for they can see only by the aid of the spiritual element; and were it not for the portion of the latter transmitted through the polar opening, earth to spirits would be involved in perpetual gloom.

We had now passed as far north as Labrador, and beneath us spread the snow-fields, the frozen ocean, and the monarch icebergs; and the terrible Odin drove across the frigid waste in biting gusts, lighting the dancing flames of his northern watchfire. We arose a distance of about sixty miles, and allowing ourselves to be borne gently onward by the flow of the tidal current, alighted on the surface of the spirit sphere.

O, what magnificence of scenery — what splendor of coloring! Words are insipid and meaningless, and the pencil would fall from the hand of the disheartened artist. In front of us was a gentle elevation, beyond which spread the waves of a blue and boundless ocean, ruffled by the slightest breath. The sky was a liquid cerulean, in which floated great island masses of clouds, like folds of silver, bordered with purple and gold. The sun was declining in the west, drawing around him his crimson cloud mantle, and blushing the landscape with his golden hue. On earth, winter had not left his stronghold, and a few daring spring flowers by the side of the snowbank alone harbingered the coming spring. Here perpetual spring breathed mild fragrance on the ambrosial air, and nurtured the flowers in beauty. The zephyrs came in invigorating breaths, scarcely stirring the delicate foliage of the palm, laden with the odors of a thousand flowers, and bearing the songs of sweet-throated warblers, chanting in irrepressible joy in every tree.

On the eminence stood a mansion, combining the elegance and delicacy of the Oriental with the solidity, grandeur, and effect of the Grecian style. Its base was a truncated pyramid of steps, on which arose elegant carved columns, entirely surrounding the building, and supporting a crystal dome. It was a vast structure, and was discernible from a great distance. As we approached it, I observed that it stood on the shore of an arm of the sea, and commanded a prospect unrivalled in grandeur and beauty. It was surrounded with lofty trees, some loaded with blossoms, others with ripened fruit; and gorgeous flowers diffused the sweetest perfume. The leaves of an iris, by the foot of the steps, appeared to be cut from emerald, while its flowers seemed carved from cerulean. A rose, by its side, appeared to be formed of exquisitely cut rubies.

"This is my home," said my spirit guide; "here, with others who are congenial in tastes and desires, I pass my time in study, in writing, or conversation."

"There are few persons here at present," I observed.

"They are away; some on missions of benevolence to lower circles, endeavoring to reform the erring and elevate the depressed; others travelling across the vast oceans of space to other worlds, observing the various manifestations of Nature; while others, still, are visiting other societies."

We entered the halls of the temple — passed the massive carved portal, and through long corridors hung with exquisite paintings of landscape. Scenes in the spirit land, on other globes, on earth, — all the interesting localities were represented; and interspersed with them were portraits of great men, among which was a delineation of Christ, said to have been made five hundred years ago. Other halls had shelves piled with specimens from all the kingdoms of Nature, where the student might retire, and by comparing her endless diversity of forms, seek to develop the great laws of creation. It was the home of a great family, who, with pure and trusting hearts, dwelt in harmony, possessing it in common, and devoting it to a common use.

As we entered one of these halls, the mate of my guide arose and embraced him. She was listening to the narrative of a noted traveller, who had just returned from a long voyage of discovery to a remote star-cluster. After they had exchanged a few remarks, the guide turned to me, and inquired,—

"Are you not fatigued?"

"Yes," I replied; "I have felt a sensation of weariness for a considerable time."

"Then you must not remain in this state a moment longer. Retrace this line of spiritual matter, which, you observe, has remained unbroken."

It was with deepest reluctance that I left him on the brow of the spirit zone; but fate, stern and inexorable, compelled me to do so, and the next moment I was again clothed in my mantle of flesh, awaking with a dreamy unconsciousness — a dim, undefined recollection of the scenes of the two preceding hours. The gloom of twilight mantled the external world, strangely contrasting with the ethereality of the region I had left.

CHAPTER XVII.

PHILOSOPHY OF THE SPIRIT WORLD.

The Spiritual Body. — Spirit Life. — OF OUR HOME. — Biblical Account of Heaven. — The Law. — Clairvoyant Testimony on Emanations. — The Spirit World. How derived. — Illustrations. — The Spheres. — Distance from the Earth. — Size. — Rotation of. — Relation of to Spirits. — How reached. — Size of the Sixth Sphere, or Zone, estimated. — Arguments against the Existence of such Zones refuted. — Circles and Societies explained. — Cause of Confusion. — The Home of the Blessed. — The Home of the Impure, (Evil?) — Relation of Spirit to Physical Matter. — How Spirits travel through Space. — Annihilation of Spirits. — Description of the Second Sphere. — Dwellings, Animals, Manners, &c. — The Society.

In the chapter on the origin of spirit, the idea was advanced, and attempted to be sustained, that the spirit possessed a material body. This body is composed of atoms, as the physical body is made up of atoms, only they are more refined. It is also made up of organs, for if originating with the physical body, and, as it were, cast in the same mould, it must of necessity retain the same form. A spirit cannot be a "detached intellect," wandering formless and shadowy. Mind, the essence of the spirit, cannot be detached from spiritual matter.

This is supported by clairvoyance and by science. It is a well-known and puzzling fact, that when a limb is lost by amputation, or otherwise, sensations are carried from it to the brain. Thus, if a leg is lost, the sensation of coldness or heat of the foot will often be experienced. Clairvoyants see the spiritual member, whole and perfect, occupying the place of the lost physical part. Hence facts precisely coincide. What is still more convincing is the impossibility of conceiving of existence without a material garb. This evidence has been before presented, and in its full bearing is incontrovertible.

The spirit must have a material garb, or instrument, through which to manifest itself. Without it, a formless, "detached intellect," it would be without power, and void of existence.

If it has a form it must require sustenance. Motion, action, thought, all necessitate waste; and if the spirit has organs of assimilation, they were made for the supplying of that waste. Nature admits of no extravagance, nor waste. Nothing, from least to greatest, is made in vain.

If subject to periods of activity, rest is implied. The mind wearies and cloys of one pursuit, and desires another, or a suspension of all pursuits for a time, until by the normal process of repose its faculties are renewed. If you imagine earth-life transposed to the spheres, you will have a true picture of our home. Each individual takes with him all his desires, knowledge, and emotions, and is the same there as on earth.

Of Our Home.

If thus materially fashioned, and holding such *physical* relations to spiritual matter, and that form of matter experienced by man being as nothing to us, we must have a home of our own. The countless millions of earth's departed cannot reside on its surface · a better and wider clime is prepared for them.

The biblical account of the spheres is inaccurate

uncertain, and insufficient. The following calculation from its texts has appeared in many of the religious papers: —

"And he measured the city with the reed, twelve thousand furlongs. The length, the breadth, and the height of it are equal. Rev. xxi. 16.

"Twelve thousand furlongs — 7,920,000 feet, being cubed, is 496,793,088,000,000,000,000 cubic feet. Half of this we will reserve for the throne of God and court of heaven, and half the balance for streets, leaving a remainder of 124,198,272,000,000,000,000 cubic feet. Divide this by 4066, the cubical feet in a room 16 feet square and 16 feet high, and there will be 30,321,843,750,000,000 rooms.

"We will now suppose that the world always did, and always will, contain 900,000,000 inhabitants, and that a generation lasts 33 years and 4 months, making 2,700,000,000 every century, and that the world will stand 100,000 years, making in all 270,000.000,000,000 inhabitants. Then suppose there were a hundred such worlds equal to this in number of inhabitants and duration of years, making a total of 27,000,000,000,000,000 persons; then there would be a room 16 feet square for each person, and yet there would be room."

Whoever the author of this sublime nonsense of mathematics may be, he has exhibited the folly and ignorance of the day. Is humanity to be thrust into such a dovecot of a heaven? Are we to be incarcerated for eternity in such a gigantic beecomb? Every rational sense forbids. Such is the church-view of the future life! How degrading, how puerile, how unmanly! Let the waters of Lethe close over the soul forever, let oblivion's wing nestle it, rather than a spiritual existence in such a place. The streets of gold, and throne of God covered with precious stones! What a show of learning! how little sense! Contemplate the Milky Way. Every sweep of the telescope brings thousands and thousands of suns to view, each having its fleet of attendant worlds. If each of the worlds which flash through the crystal vault of night were to send a single delegate to the throne of God, this heaven would overflow, being packed to its utmost capacity!

Such a heaven would be the grand miracle of creation, such as an Oriental despot would build could he possess Aladin's lamp, and have all his desires gratified by the discovery.

It is not the sage's heaven, nor that of the rational man, any more than is the sensual paradise of Mohammed.

Here we present the great proposition — one which will awake the world from its lethargy, and make the fearful quake.

IF THERE IS A HEAVEN — A HOME FOR THE SPIRIT — IT MUST BE ORIGINATED AND SUSTAINED BY NATURAL LAWS.

We have shown how laws and principles support the physical world.* We have proved that they regulate the world of mind. We have proved that miracles, the suspension or overruling of a law of Nature, are impossible, — these laws forming a part, being an integral portion, of matter itself; also, that spirit is matter, and governed by law. If so, then, how avoid the conclusion that the spirit spheres are ruled by law?

* In the previous volume.

Such are the arguments we present to the reflecting mind to corroborate what we shall hereafter communicate. What we have seen we know, and the evidence of our senses, whether received by man or not, is as positive evidence as any thing that can be educed. Such evidence we shall now bring forward.

Matter is subject to eternal progress, from the granite rock which juts to the sky in the craggy mountain peaks to the atoms of blood coursing through the veins of man. Matter arises from the crude angular to the refined spherical. Still further is this process carried in spirit, which is sublimated matter. From all worlds the latter ascends as it is freed by the processes of life. We can see it escaping from the rock within which chemical forces are at work; from the growing or decaying plant, set free by light; and from the dying animal like a vapor.

[While in a state of clairvoyance, I beheld this process in a most beautiful manner. I was seated on the brink of a limestone cliff skirting the shore of Kelley's Island. The waves of Lake Erie dashed gently at my feet. I had been writing by impression on this subject, and the influence which impressed me I supposed had withdrawn, when suddenly I became clairvoyant. The waves became irridescent with the blended hues of a myriad rainbows; but this soon vanished, and then I saw what Reichenbach would call the odyle of the waves ascending and enveloping me. But I have another name and explanation for it. I saw that it was a spiritual emanation, and originated from the agitation of the water and decomposition of dissolved organic and inorganic matter. I could feel the presence of this emanation to a considerable distance from the shore, especially when the wind blew over the water, even in my normal state, but could see it only clairvoyantly. It then appeared as a delicately-tinted ultra marine, greatly rarefied, gas. When it arose and flowed over the edge of the cliff, like a beautiful cascade, directly upon me, it produced the most delightful sensations I ever experienced.

I now recall to mind the great irritability produced on my nerves by the sea breeze, which I still vividly remember, though I was then a mere child, and wholly unacquainted with the cause.

After I had this vision of the sublimate arising from the waters, I was spiritually transported to the side of a dying animal. The blood had already stopped circulating in its veins; all the vital functions were still. The process already described as occurring at the death of man I saw taking place; but when the vapory cloud arose above the body, and the connecting cord was broken, the cloud, instead of reverting to the form of the animal from which it arose, — as I had repeatedly seen it revert to the human form over the corpse of man, — evaporated before me, and mingled in the ascending current of heterogeneous spiritual sublimate.]

This spiritual *substance* is an advanced stage of development of gross matter, and is attained by the principles of progress inherent in the ultimate molecules of which matter is composed.

Thus derived, we have but to follow its course to know what becomes of it, and what offices it fulfils. If we do so we shall be carried in a slightly spiral line through the polar opening, and find ourselves in the second sphere. Then we shall see these currents dispersed. They go there to build the second sphere — the home of spirits. Earth not only gives existence to identified spirits, but also to non-identified, which build the home of the former. How and where we will now determine.

The second sphere surrounds the earth like a very broad belt, extending sixty degrees each side of the equator. Hence sixty degrees are left unoccupied at each pole, which explains the term "polar opening," previously used. The position of the spheres may be better understood by reference to the engraving.

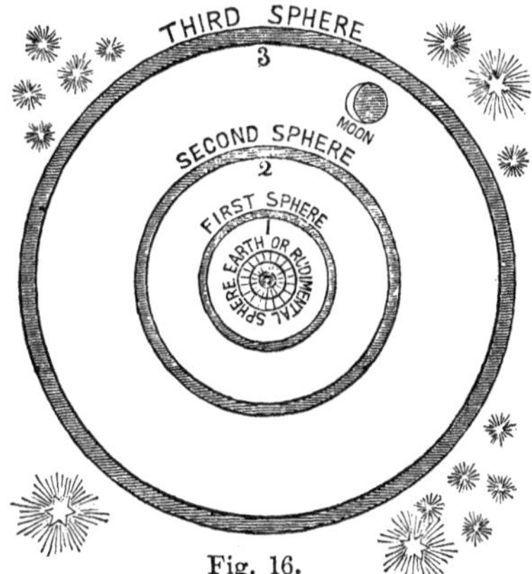

Fig. 16.

The second sphere is about sixty miles distant from the earth's surface, and nearly fifty miles in thickness. It is a solid belt or zone. We say solid, for it is so to us, or holds the same relations to us as solid bodies do to man. It is constantly increasing by additions of new material brought from the earth by ascending currents. *Substance* — a word we use to describe spiritual matter — arises to the level where its repulsive and attractive tendencies are equal, which is the position of the second sphere. Then its ascent is arrested, and it enters at once into this formation. Still subject to planetary laws, it rotates around the earth as its centre. The zone rotates, and hence its plane of rotation must coincide with that of the earth. Its period of rotation around the earth is a little more than twenty-four hours. Only its surface is inhabited by spirits, as man inhabits only the surface of the earth.

In the second sphere the same refining process goes on as in the rudimental or earth sphere. Matter, even there, is far from its ultimate. Three spheres extend in a similar manner around the earth, the most external lying beyond the moon, that luminary revolving between the second and third spheres. The size of the second sphere, or the extent of its surface, is almost incomprehensible. If we consider it situated at the distance of the moon, and do not make allowance for the polar openings, by mathematics it is determined that it has a surface of 732,601,572,000 square miles, or is equal to 305,291,723 worlds the size of earth, provided their surfaces are entirely inhabitable. It will be perceived that the problem so perplexing, If man is immortal, where in creation, however vast, will room be found for so many spirits? is solved in the most beautiful manner. The same process produces the spirit and the material of its home.

If there are spheres around the world, why do they not interfere with the light of the stars? The atmosphere is a dense gas, yet it freely admits the rays of light, and is perfectly invisible. A sphere of crystal could not be seen even. Yet how coarse are these to the agents we are considering! How, then, can they obstruct the rays of light? or how be seen by man? Hence, because they are not seen is no proof that they do not exist.

The division into spheres is a real division, but that of circles and societies is purely arbitrary. It is from this cause so much confusion exists in descriptions of the spirit land. But in this confusion there is no real contradiction. The numbers of societies and circles are infinite, as they are on earth; but the better to describe the inhabitants of the spheres, and their manners and conditions, this classification was introduced.

Seven Spheres.	Six Circles in each Sphere.	Three Societies in a Circle.

This should be written Three Spheres, number of circles and societies in each infinite. But the number of circles and societies is wholly arbitrary, no lines existing more than on earth. There you perceive congenial minds are drawn together into groups and coteries. Such are effects of the law which here supremely rules.

It would be a useless task to describe the second sphere, much more the ineffable glories of the superior, which we have not seen, nor expect to see for a million of ages. We cannot paint the ecstatic beauty of the scenery, the clearness of the sky, the azure of the ocean, the beauty and depth of color of trees and flowers. Give wildest imagination freedom to picture all that is desirable and beautiful, splendid and glorious, yet no idea can be formed of our home.

Pausing a moment, we will examine the apparent confliction of this division into spheres, with the other division into seven. Each planet has spheres, in size, rotation, and number as diversified as their satellites. The earth, regarded separately, has three; but the whole solar system, as one, conjoins to throw around itself, as a whole, spiritual zones held in common, as they encircle the solar system. And, carrying the analogy farther, this entire world-continent, the vast galaxy, — or what is usually called the universe, — throws around itself, by emanation from all its component solar and stellar systems, a series of ascending zones. These zones unite and blend the universe into one whole, permeable throughout its whole extent for refined spiritual beings.

The planets fade and melt into their second spheres. These, in their turn, fade into the third; these ascend into the solar or celestial spheres, four in number; these successively ascend, and at length merge into the infinite supercelestial spheres in which the universe is encircled.

From such speculations — you may so call them — we turn to the more practical part.

Such is the home of pure and noble spirits. Where is that of the vile and debased? On earth. You have a world, an invisible world of spirits, constantly around you. You see them not; you neither hear nor feel them; yet are they with you, dwelling by your side. Such spirits are chained to earth by attraction, and cannot rise into the second sphere until they have cast off the taint of earth. These are the spirits who, in their communications, tell you that animals exist in the spirit world. To them they really do. Spirits cannot discern physical matter, more than man with natural vision can see spiritual matter. When they look at an animal, they cannot see its gross form; they can only see its spirit: they cannot see the external plant; they see the internal plant: it is not the external, but the internal, world which meets their perceptions. Earth being *their sphere*, they consider the spirit world as not only inhabited by spirits of man, but of beasts also; and they also say truly that the latter are necessary for their existence; for when the animal's spirit dissolves at the death of their physical bodies, they eagerly absorb the dissipating vapor.

Clairvoyance affords proof that spirits can only see spirit, for it is often with difficulty that the clairvoyant can distinguish between men and spirits, the spiritual portion of the former only being perceptible; and it is not by aid of simple perception that the distinction is determined.

We say that such spirits are chained to earth by attraction. This leads us to an explanation of how we travel through space.

Our bodies are more refined than the ether of space, and hence our gravity is entirely destroyed. We float in this medium, and our course is directed and impelled by our will. If we wish to go to a certain place, the attraction created by our will is sufficient to impel us to that object with almost the rapidity of thought. But when we rove among the scenery of earth, we walk, as we were wont when mortal, as it brings to our minds the visions of that life.

Not so with the earth-chained spirits. They are denser than the ether of space, and hence cannot rise to the second sphere more than man with his gross body. The surface of the earth is their home. The Indian spirit builds his wigwam on the high bluff, overlooking some beautiful streamlet or lake; the bacchanal makes his haunt in the den of corruption; each finds employment — each finds an appropriate sphere. Each is happy in proportion to capacity; but, O, it is a miserable state.

How many expire like the ill-trimmed taper! Others advance, and, shaking the leaden dust of earth from their garments, ascend to the spheres. With some this is rapidly effected, with others it is the work of ages.

The second sphere is a daguerreotype of earth. The refined matter which ascends is prone to assume the forms from which it was liberated on earth. The scenery is identical, but more beautiful and ethereal. Trees, fruits, and flowers are not individualized; that is, their emanations do not ascend in an identified form, but particles thus emanated are more prone to assume those forms than any other. Thus the particles which exist in a particular flower have never existed in that flower before, but have ascended from a countless number of flowers of the same kind. The description of the splendid scenery, the gorgeous landscapes, we reserve for a future chapter. One thing only remains for us to elucidate. We speak of dwellings — of artificial things — as existing in the spirit world. Are they created by our simply desiring them? So, many spirits have falsely taught. It is true, our desires create them; but we employ means, just as man does, to accomplish our wishes. We are not miniature gods, capable of creating a palace by a wish. The marvellous powers of Aladin's lamp are denied us.

This is true of the lowest and the highest spirits; and in this respect none are superior to man.

If we desire to unite in societies or groups we do so, and we find our happiness is increased, our enjoyments doubled, our progress accelerated by such unions of love, in which perfect harmony reigns. If we desire instruments, machines, dwellings, the unassisted desire never can obtain them. There is wood in the forests, metal in the mountains. We have the means to incarnate our desires.

Such is the spirit world. It is a world. It is a matter-of-fact world, more real than earth. It is no ghost land — no vale of shadows — but the ultimate essence of reality.

CHAPTER XVIII.

SPIRIT LIFE.

Office of Spirit Revelations. — Their Necessity. — Spirits retain all their Faculties. — Affinity. — Condition of Good and Bad. — No line of Distinction drawn between them. — Condition of Spirits. — Rewards and Punishments. — The Miser. — The Animal Man. — The new-born Spirit an exact Copy of the Man. — A more cheering Picture. — Unchangeable Fiat of Organization. — Capabilities of Spirit. — No Forgiveness. — Earth a Primary School. — Better Conditions in the Spheres than on Earth. — Spirit Missionaries. — Their Labors. — Heaven of the Astronomer; of the Philosopher; of the Poet; of the Historian, &c. — Conjugal Love and the Marriage Relation. — Retention of the Animal Faculties. — Difficulty of elevated Spirits communicating with Earth. — Lower Spirits can more readily do so. — Why? — The Doom of the Suicide. — Heaven and Hell. — Conditions of Mind, and on Earth as much as in the Spheres. — Life of a True Philosopher.

This subject has engaged the strongest minds, in vain effort to pierce the veil which conceals the future life from view. They have all failed, because they were imbued with a sensuous philosophy, and the spiritual essence cannot be recognized by the senses. So difficult the subject and unsatisfactory the results of thought, that many have been frightened away by the barren cheerlessness, and confirmed in the mistaken notion, that the Deity has set bounds to human reason, and it is sinful to strive to see beyond, or taste the forbidden fruit of immortal knowledge; that he should only attend to things of the earth life, not endeavor to look beyond.

But spiritual revelations have set man aright, and taught him that the earth life is nothing, compared with future glories. Those who have passed from earth, return with glad tidings to cheer their desponding friends. We tell you of the spirit land as travellers describe the countries they have visited. Say not that man does not desire this knowledge. It is an instinct of mind, and eagerly craves gratification. The traveller desires to learn all that is known of the country he is about visiting; and if an unknown, unexplored region, how eagerly he grasps every item of knowledge concerning its inhabitants and climate, that he may prepare himself for the journey! The Hebrew must send spies ahead to determine the character of the Land of Promise.

Do not turn poor striving humanity aside with the cold-hearted answer that they must not enter the unknown; that the future is not for their ken. Do not say that such knowledge is unpractical, and of no consequence. It *is* practical, and of utmost consequence for man to understand the future life, and whether that life is divided into a golden heaven and a fiery hell, as taught by the popular theology.

It is the received belief that a strong line of demarkation exists between man and spirit; that the spirit at death enters a new sphere, where every thing is changed, and the laws of Nature act differently from what they do on earth. Indeed, the change is so great that identity might as well be lost, for in the spirit the man cannot be recognized. It is taught that the moment a person dies, if he has acted well, he knows more than all the living; but if has done badly, endless perdition is his. Thus a sharp line is drawn between the good and the bad. But the division is false. There is just such a division as on earth. He who places himself in inharmonious relations to the laws of his being, is in hell, or misery, until he places himself in harmony with them. There is a retributive recoil to law. The suffering, punishment, is in direct ratio to the extent of the inharmonious relation. Great or small, it is the same; the spring recoils with precisely the force with which it is bent. Do a little wrong, and the recoil is slight; do a great wrong, and its force is proportional. After death this is equally true. The spirit holds the same relation to spiritual matter as man holds to physical nature. As spiritual matter possesses all the properties of form, extension, impenetrability, that physical matter does to man, hence the spiritual life is but an *extension* of man's existence, enjoyed amid pleasant scenes and sublime beauties, surrounded by Nature in her most magnificent forms.

It belongs to the wild vagaries, which, like mushrooms, have grown from the waste material, the accumulated rubbish of thought, to write its own history of errors. We shall not fatigue the reader with a history of beliefs, nor pause to write an epitaph over the wrecks of their folly, but refer them all to the active but encumbered intellect.

Mankind are strongly prejudiced against crime, and desire to see it punished. Sin must meet a terrible retribution; if not on earth, then in the next life. For a moment lull passion to sleep, and awake benevolence and compassion. It should be ever remembered that Nature inflicts suffering, or what is called punishment — not for revenge, or because she delights in pain — but for the resulting good. To those, however, who are constituted to believe in a hell and heaven, the idea of good and bad going to the same sphere is intolerable. They cannot see how good and bad can exist side by side, without an impassable barrièr between them.

But see how it is on earth. The rains water the fields of the unjust as well as just; the dews fall, and Nature toils for the debased and criminal, as well as for the good. She treats all her children with impartial hand. She says, I treat you all alike; now do your best.

Yet every one receives his deserts. If all the inhabitants of the earth were translated to-day, they would hold the same relations to each other in spirit life as at present. Previously to the change, no separation is needed, nor is any thereafter. Affinity controls the relations of man, and he forms such relations and connections as are congenial, and avoids the opposite. The low congregate together, and avoid the society of the refined, while the learned and moral delight in each other's society.

Though there is no artificial barrier between the two classes, there is a natural separation, wide, deep, and impassable as the fabled gulf between hell and heaven. This is the affinity and repulsion which exists between differently constituted minds. This is seen even in animals, as has been already shown. Different herds will not mingle, and when forced together, soon separate into the original flocks. Men separate in a similar manner. Those who delight in each other's society congregate together. Learned societies, scattered throughout the world, draw each an appropriate class of minds. Antiquarians meet to discuss the misty past; naturalists to compare specimens; philosophers to exchange theories and facts;

astronomers to compare observations; linguists to discuss languages; statesmen to plan schemes of government. Thus each finds congeniality, and enjoys the society he chooses above all others. Are these bodies disturbed in their deliberations by the idle curiosity of ignorance? Do the debased resort to their halls? If so, they soon go away in disgust, to the theatre, the race course, the gambling saloon, and dens of infamy. When they meet those who sympathize with them, are they molested by the refined? They fear not the intrusion of those who would be disgusted with their brutality.

There is no necessity of a gulf between these; they cannot intermingle.

The spirit is the representative of the man, retaining all his thoughts, ideas, and impressions identically the same. Then there is no more need of an arbitrary division of spirits. The just and the unjust, the good and the bad, the debased and the elevated, the degraded and the virtuous, all *enter* one sphere. Ah, but do they not all *leave* one sphere, in which they dwelt side by side?

We recognize neither hell nor heaven; we can see only happiness, with an alloy of pain. Joy is painted on every leaf and flower, breathes in every zephyr, flashes from every star, fills every warbler's throat with richest music of love, breathes from all nature. Mingled with this redundancy of happiness are channels of misery; not by mistake, nor cruelty, nor injustice, but from motives of love, justice, and benevolence. Man is imperfect; the earth is imperfect; but both are perfect compared with each other. If man were perfect, he could not exist in so imperfect a world; if the world were perfect, so imperfect a being as man could not exist on it. They are mutually adapted for each other, and equally progressive; they move onward towards perfection together.

Thus it is in the immortal land. There is no sharp line of distinction. In the universal justice which prevails, blessings fall alike on all, according to their capacities. Not that misery does not exist. Wherever there are fallible and imperfect beings, misery must exist. It is the result and constant attendant of an undeveloped state. But there is never more than is absolutely necessary. It never exists for its own sake.

All are as happy as they are capable of being. The man of great and lofty thoughts looks down on his brothers, striving and wearying with petty difficulties and trifles, and from his heart pities their contracted souls, to be filled with such nothings. The angel looks down from the immeasurable height of his star-lit home, and perceives this exalted mind far lower in his scale of comparison than *he* saw his earthly brothers. And angels far, far above, look down from the celestial spheres, and consider the lot of this bright angel sad, his cup of happiness not half full. Thus all ideas are comparative; but the causes of happiness and misery are alike with men and spirits. Happiness is the result of harmony; misery, of inharmonious relations.

It is true, many times death makes the spirit more miserable than the man. Wherever enjoyment is derived from the animal, instead of the rational faculties, this results.

For instance, the miser has little enjoyment on

earth, and that of a purely sensuous character. His soul is contracted and dwarfed by his occupation of shuffling the cards of speculation, and hoarding his profits. Did we say profits — moneys? Rather concrete sighs and tears of the poor, the wretched, and despised. Every avenue through which true happiness or rational enjoyment can enter his mind is closed. Only one remains open, and that a stagnant sewer of corruption.

Suppose that he is transplanted into the spirit world. He will remain the same individual; not in the least changed; with all his desires as ardent as ever. But if *he* has not changed, the *circumstances* which surround him have entirely. He is ushered into a world of freedom, where all have as much as they need, and the damning doctrine of mine and thine is cast contemptuously aside. He enters that sphere stripped of all externals; gold, station, honors, all are left with the body. The mind stands there alone. That mind so neglected and abused, over which so many storms have swept, resembles the old blasted storm-torn oak — a miserable wreck of manhood. It is a fearful crime to sacrifice the man on the altar of insane desire for wealth, and it meets a severe punishment. Can you ask if he is happy? Can he be otherwise than wretched? The beauties of the angel land are not to his taste, and he immediately gravitates to the debased societies, where he meets others like himself, where together they can reenact the diabolic scenes through which they have passed on earth.

Here is the animal man. He works and eats; awakes to work, works to eat, eats to work again, sleeps to eat and work again. With a brute eye he gazes on the sparkling orbs which hang their signal fires in the sky, making the arched blue redolent with diamond glory. He thinks the flowers pretty, the sunlight good, but he never feels a deep yearning aftei a high, holy life — after pure, spiritual thought.

He enjoys the delights of the animal, and no more. He enjoys working, eating, and sleeping. These are what he lives for, and all his happiness is derived through these.

In the great beyond all these relations are changed. Animal life in a great measure is laid aside. The weary soul, when wafted over the Styx, is stripped by relentless Charon of all titles and conditions which minister to the pleasure of the animal man. There it stands on the borders of eternity, with scarcely the capacity of an infant; its faculties half obliterated by the cares of the world; its intellect dwarfed: without an accumulated store of wisdom, it finds little to cheer, much to perplex, and passion, which has never been controlled, sweeps with destructive violence over a mind bleak, rugged, and desolate, as a granite peak of a snow-clad mountain.

The low will seek the low, mingling with comrades with whom they affinitize. They will enjoy themselves as best they can; engage in conversation on grovelling subjects, or wander over the world in search of treasures, or plan useless schemes of speculation.

We here present a fundamental truth: —

The New-Born Spirit is an Exact Copy of the Man. If ignorant when he departs earth life, he will be ignorant still; if wise, then will he be equally wise. Nothin is gained, nothing lost.

[To illustrate this primary fact, I will introduce an incident drawn from my own experience.

Quite a large circle had convened for the purpose of obtaining communications, at the residence of Mr. J——. They had been entertained with a speech, made by an ancient spirit, and were waiting for further manifestations, when an Indian chieftain took possession of Mr. J——, and began to complain of the wrongs he had suffered, and that his race were still suffering. As he continued he became angry, raved like an imprisoned tiger, and spoke so loud that it was painful to those near him. The spirit had frequently taken possession of the medium before; but he always spoke of his wrongs, and became so impatient that the medium could not speak in public. The spirit gave evidence of a fiery, savage state of mind, desiring to tear and crush every thing before it, rending the medium, tearing his hair, &c.

After the first outburst of passion had partially subsided, Mr. T—— became influenced by the spirit, who had previously spoken through him, and exclaimed, —

"Indian friend, do you not know that it is wrong for you to indulge in such storms of passion? The past is past, and you do not benefit yourself by mourning or becoming angry over it. You have suffered great wrongs. I read them in your mind. You have been trodden beneath the foot of tyranny, dispossessed of your lands, and exterminated; but you do nothing towards obtaining justice by such outbursts of anger." The Indian replied in a low and conciliatory tone, "Yes, I do wrong! How sweet the words of sympathy! They are the first I have heard for many long years. Yes, I do wrong. But how can I help it? Driven from the graves of my fathers by the ruthless assassin, my wigwam burned, my pappoose killed, my relatives slaughtered, and left for a prey to the wolves, — my blood burns; I want to fight, to kill; give me my tomahawk, or my knife and rifle; I'd murder, scalp, and burn." Again he became angry, and raved. After he had subsided, the ancient spirit replied, —

"This is entirely a wrong spirit to indulge; you should not look at the past, but at the future. The past is dead; the future is your own. An immortal spirit should not thus indulge his animal nature."

"I know, I know; yet how can I help it? I used to hunt on these grounds; many a time I've chased the deer over these fields, and my wigwam was close by. Now the red man has perished; his wigwam is gone, the forest has melted away, and a smooth ploughed field my hunting ground."

"But you must remember that such is the destiny of races. The Great Spirit has given a title deed of lands to that race who uses it the best. A thousand whites live on the same ground that you occupied. You lived on game, they till the soil; and every one of this thousand enjoys more than you did. You see the thousand have the right to till the land which the one desires to be a waste; and as you would not cultivate it with them, they dispossessed you. Such is the law of races, and you must abide by its justice. But this is not your present concern, which is to improve yourself."

"I cannot improve; I cannot take the first step; I am doubtful whether I can ever overcome my vindictive passions."

"You see that bright spirit there, enveloped in a garment of silver so brilliant that you can scarcely bear the light of it for a moment, whose gigantic mind understands the profoundest problems of science, and whose morality is godlike. That spirit was once lower than you. Once he wallowed in the lowest depths of sensuality and crime; but a benevolent spirit descended into his low sphere, and pointed him upward, and ever since he has been advancing. Look; do you not believe that you can become as pure and holy?"

"I cannot believe."

"Will you try?"

The Indian remained silent, looking thoughtful and downcast.

"Put your foot on the first round of the ladder, and climb upwards."

"Nay, I cannot. It is useless for me to make any exertion. I must remain where I am forever. If I should make the effort, I am surrounded by such associates that I should soon fall back into my present condition." He sank back with a despairing and hopeless expression, while tears fell fast from his eyes.

"True, but I dwell in a temple surrounded by every thing that is pleasant, beautiful, and pure; the groves resound with the love songs of the warbler; the ocean sighs at its foot, and an azure sky bends above it; there none but the good, wise, and loving dwell, who will sympathize with you, and endeavor, by every means in their power, to elevate and teach you lessons of morality: there no disturbing influence will counteract your efforts to improve yourself, but all will assist and encourage. You cannot advance as long as you remain with your present associates; you must breathe a higher and purer atmosphere, and be surrounded by purer minds. These you will find in our circle, where you can advance as fast and as far as you have courage to desire. Will you go with me, and become a member of our association?"

The Indian was silent, then fell on his knees, and grasped the hand of the spirit.

"Go!" he exclaimed; "may I go? Then I am blessed. Never have I had sympathy before. It has touched my heart; I will go any where with you."

"You have taken the first step; now climb. Look not down, but upward. Keep your eye steadfastly fixed on the sun of righteousness, and swerve not a hair's breath. Come, let us away." He grasped the Indian's hand, and the influence immediately left the medium.

Two weeks after this little incident, as interesting as it was unexpected, the Indian chief again influenced Mr. J——. He was not angry, tiger-like, and thirsting for blood, but calm and loving. He said that he dwelt in the society of the philosopher who first became interested in him; that his associates instructed him; and under the beneficent influence they shed around him, the clouds were rapidly rolling from his mind, and he was entering a pure and heavenly existence.]

We can present a more cheering picture of woe and suffering. Those who cultivate their minds reap a rich harvest. They have laid up a treasure in heaven! Charon wafts this over — THE WISDOM, LOVE, AND PURITY OF THE MIND. Whatever is known to the man will be known by the spirit. Knowledge is all the spirit carries with it. All else is dross. When the man of thought awakes from the sleep of death, he finds the avenues through which he derived *his* enjoyment open wider than before, and at once feels the breath of heaven above and around him; resplendent beings welcome him to their gorgeous home, where he can pursue his studies untrammelled. How ecstatic the delight of such a spirit, when such pleasures touch its tender chords! What splendid visions of thought! He finds sympathy and congeniality. All his faculties are exercised. He is surrounded by *three* conditions requisite to make heaven for him, and fully gratified, he at once enters into its joys.

All denominations and sects, in their rational moments, when not maddened and blinded by the lust of controversy, have distinguished between the future states of the depraved and the good, between the ignorant and the cultivated, by saying that every one's cup is filled to the brim — if it holds a pint or a gallon. But they have harnessed a great error with this truth, by denying progress, denying repentance after death, by saying the cup would always remain of the same capacity.

From necessity of organization, a Hottentot or Bushman, the barbarous savage of the wild, can enjoy little on earth — still less in the angel sphere. While the philosopher studies the mysteries of his new home, wanders from orb to orb across the vast oceanic spaces which separate the world atoms of the universe, approaching nearer and nearer infinite comprehension, basking in the sunlight of knowledge, the savage must grope in darkness, not knowing where he is, or what he must do to exalt himself from his grovelling position. Such are the extremes we often meet.

Some of these dim lights expire; but if endowed with the feeblest spark of progress, the lowest can in the ages of eternity equal the highest. Many a debased savage will in the distant future arise to a plane of development which we cannot now comprehend. The criminal, in whom conscience seems blighted and crushed, will surpass the piety of Channing. The miserable beggar will become a more profound philosopher than Descartes. Such is the consoling doctrine of progress. We should pity these outcasts, while we recognize the presence of an immortal spirit germ, capable of infinite unfolding. A diamond is concealed beneath the reeking rubbish, which one day will burst forth in brilliancy. The rudiments of a mind exist, which, when placed in the proper conditions, will surpass the piety of a Paul, the philosophy of a Herschel, the grasp of thought of a Humboldt. This is the doctrine of reason, based on the progressive nature of man.

Men are not alike — nor can they ever be. Conditions surrounding them will stamp their character indelibly. Yet all can be good and great. The brutal men of the world are results of brutal conditions, and by making the conditions harmonious a millenium will be ushered in. Each individual has a certain bent, from circumstances, and this is called *genius*.

Men never cultivate all their faculties. One philosophizes at the expense of his moral and social natures; another moralizes at the expense of his intellect; none cultivate all these equally. This is wrong. The constant use of one set of faculties, and the inactivity of the others, induces disease. As the mind holds an intimate relation to the body, any mental disturbance equally affects the physical system; and the excessive activity of one portion of the brain, and inactivity of the remaining portion, induce inharmony, and thereby bodily disease and intellectual eccentricity. Why are there such multitudes of wrecks of humanity bragging their wretched forms through life? Simply because the law of harmonious development has not been heeded. Infinite time only can complete this harmony. The man who cultivates his whole being, and not a part to the neglect of the remainder, may not flash out in an unexplored field beyond the extreme outposts of his daring predecessors, but he will be founded on a solid base, and in no danger of a fall. The mind should grow as a tree, each year adding a new circle to its former limits. Then it will be always prepared for the change.

Such should be the training of the child; but when mature, harmony can be restored. Judgment is far too hastily pronounced on the poor, debased spirit, ruined beneath its sins, that it can never rise above its brutal state. This decision is drawn by comparative reasoning. As the young tree is bent, so it remains, growing more and more irretrievably into its distorted form. Its distortion is eternal. Not so with the spirit, between which and the tree no analogy exists. As stern and iron as mind appears, it is plastic and yielding. Like a fluid, it adapts itself to the circumstances in which it is placed. When circumstances are unfavorable, it remains dormant, or is dwarfed; but as soon as they become favorable, it overcomes the deformities of the past by vigorous growth.

There is no forgiveness. Law pays not the least regard to prayers or intercessions. Do wrong — i. e.. become inharmonious with yourself or with Nature — and you will be inevitably punished. Much or little, it is the same. You cannot escape until the utmost farthing is paid. The only forgiveness of sin is the punishment. When you have endured that, you can go on fresh and new, having lost just so much, however, of what you might have been.

If you lose a limb, is it ever restored? If you are burned, does not pain follow? Is not a scar left? This is a bitter proposition, but it is, nevertheless, true. The contrary doctrine of atonement is ruinous belief, which, no one in his manly moments can entertain. Man cannot do wrong a lifetime, and then, by a death-bed prayer, obtained through fear, be forgiven, and enter heaven as happy as saints. This is not a doctrine to live by, however good it may appear to die by. It is neither the way to live nor to die. Such prayers, such repentance, avail nothing. The spirit retains its wounds and its stains. The wounds may heal, but ghastly scars remain to record the laceration. Man should live right on earth, and then, when death throws open the portals of the spirit world, he will but step across the threshold, from one room to another. Death does not close the period of repentance. Whenever the mind resolves to change, and thenceforth become better, more upright and manly, it can do so.

Earth is the primary school, where mind is prepared for the college of eternity. If the youth ignorant of his grammar or arithmetic should be sent to Harvard or Yale, he would find it an extremely difficult task to keep with his class. His classmates, having a basis already laid, and minds trained in the processes of thought, will learn so much faster than he who is not trained by previous study, that he will be discouraged, and his mind thrown in constant confusion.

Such is the situation of those who leave earth unprepared for the next state. They are freshmen among sophomores. They want congeniality, sympathy, guidance.

If the mind remains, external conditions vary. The slave is a slave no more. The master cannot apply the lash, or grind him beneath the brute. He becomes a free, untrammelled man. The inferior races cannot be subdued and abused. The strong cannot triumph over the weak, for right makes might, and not might right. The means of progress are thrown within the reach of all, and those who desire can advance much faster than while on earth. They can rapidly ascend from sphere to sphere, each step widening the conception of Nature and Nature's God.

The son, gone before, may become the teacher of the father who remains, and many a slave, toiling in the stagnant rice swamps, amid rank, miasmatic exhalations, will in future ages instruct his brutal master in the beautiful doctrines of love and peace. The red man, who flies before civilization towards the setting sun, has capabilities transcending mortal conception, and one day will call us all pupils to his revelations.

However high the attainments of mind, however exalted its aspirations, still higher can it attain. The wisest have a meagre knowledge. Man looks upward to the unknown with humiliation; and the angel from his star-lit home looks upward likewise, and when he beholds the wisdom beyond, the vast amount of knowledge he has acquired becomes an insignificant atom, a mote in the future effulgence of glory. Thus forever. There are no limits to the wisdom of the universe, no limits to the capabilities of the human mind for the reception of truth and the enjoyment of happiness.

We are more philanthropic than man. Many there are who adopt the transcendental precept, "Do unto others as you would have others do unto you;" and not only that, but a still higher, Do ALL FOR OTHERS, thus eradicating every vestige of selfishness. Such make the object of their existence in schemes to elevate the debased circles. They take the fallen by the hand, and call them brothers, pointing heavenward to the blessings of a better and purer life. We have often heard these bright angels speak, for hours, to a congregation of ignorant, selfish spirits — who for ages had been plunged in night — of the heaven they might enjoy. His words were the burning eloquence of heartfelt goodness. He could not appeal to their blighted reason, and hence addressed their feelings — their intuitions of right. A few were touched by his words, and, after the crowd had passed on, remained to hear more of the new truths. Then I have seen that generous spirit take them under his especial care, and devote himself entirely to them, until they became firmly founded in the truths he taught them, and were enabled to stand alone. Glorious spirit! — a conqueror greater than all the boasted heroes of the past.

There are multitudes of these Howards, who are constantly engaged in such generous toil. If not for them, the low spirits could never arise from their abject condition; for, associating only with those equally low, they never could receive additional light; on the contrary, such association only plunges them deeper in sin and crime. But the intense love for their erring brothers impels the philanthropic to force themselves into these societies, and teach them the better life they can attain if they will make the effort.

This employs a great number, but is not forced on any. Those only employ themselves thus who are compelled to do so by their love, affection, and benevolence. Each seeks an employment adapted to his capacity and taste, in which the most happiness results. The lover of beauty finds landscapes of dazzling grace, quiet bowers, grand forests, babbling brooks, where he can seat himself and dream away the hours; there the human form presents itself in purest loveliness, and painter and sculptor find real models, surpassing their earthly dreams.

The astronomer is not confined to one world, but rambles at will over the universe, exploring the worlds he dimly saw as points of light through his telescope. Then his heart beat high, and his voice was low and agitated, as he strained his vision to catch a glimpse of the sparkling point of light; now he journeys over it, and notes all the variations he observes. He can stand out and look on revolving systems projected against the blue groundwork, propped by Titan's hand, and watch their stupendous movements, and on that magnificent map draw his mathematical diagrams, and develop by calculation and observation the silent forces which control the rotating globes.

The poet from sequestered retreat weaves the subtile web of fancy, adorning the wild thought with exquisite diction. He roams in worlds of his own creation.

There he holds the mirror of his soul to nature, and paints the flitting shadows as they come and go in infinite variations. He writes not for himself alone. All the poets of the past are his fellows, each striving to excel in beauty of style, elegance of diction, and nobleness of thought; meeting in societies to cultivate their understanding.

The chroniclers of bygone ages discuss the events of the past, and draw inferences or prophesy. Those old historians of a thousand ages, through which they have lived, all that time having a deep interest in the actions of nations and races, can give lessons in philosophy deep as the oracles of the gods, and eager audiences listen to their discourse.

In such minds often originate ideas which, imputed to some impressible mind on earth, overturns all previous theories, and lights up a world-wide conflagration.

The love the spirit retains for earth never expires. It always returns to its infant home with delight; and often this love for earth is greatly detrimental to the spirit's advancement, by causing regret and repining.

Those who derive enjoyment through the intellect or morals enter at once on a broad field of happiness, and are benefited by the change. It is far otherwise with the brute laborer, the depraved, all who enjoy only by means of the body. We cannot paint the misery such suffer when they find all the channels through which their enjoyments flowed closed, and they thrown entirely on their minds for enjoyment. Addicted to habits which have deeply impressed themselves, feeling the intense cravings of these without the power to gratify them, they suffer the most excruciating tortures. Words are incapable of describing their situation or appearance,—a desert waste on which only a few rank weeds mature; a soil in which it seems truth can never vegetate! It is an awful spectacle. A godlike mind wrecked, a black and dismal wreck, is too horrible to contemplate. How can such be happy? Though surrounded by all the pleasures of the blessed, the same sun lights their way, and the same objects meet their gaze: these, instead of pleasure, from their inharmonious relations, inflict pain.

The spirit requires rest. When it is fatigued by close and unremitting study, it is necessary for it to retire, as it were, within itself, and take repose. It is not the body alone which sleeps, but the spirit also, and when freed from the body the necessity for sleep remains the same. It cannot go on in gigantic labors without intervals of rest. This gives a zest to life; for arising from repose, after fatigue, fresh and strong, adds very much to the enjoyment of living beings.

Spirits retain what are called the animal faculties. They give energy to the mind, as steam force to the engine. The man of intellect alone, though coupled with the morality of a god, is incapable of effort. He can never exert influence or power. He is discouraged by the slightest opposition, and one by one sees his most cherished plans forsaken. His voice is feeble, and silenced by the louder blasts of error, whose stentorian voice rings with the force of combative energy. So would it be with spirits. The animal faculties are not depraved more than the intellectual. Their uncontrolled action is evil. They are as necessary in their place as reason or conscience. Without them, the others become abstract ideas, incapable of application, and only through their energy are the higher faculties exerted.

Holding such relations, they should not be regarded as sinful, but as the force of character. To properly direct them should be the study of every individual. They are as necessary for the spirit as man; and from principles previously explained, they must be retained in the same relations to the whole mind as they held before death. Nor are they better controlled in the lower circles, but in the higher they are in perfect harmony with the purest morality, and send the philanthropists on grand missions of good or herculean labors for the welfare of their brothers. They send such to earth to teach man the interrelation of earth and heaven.

Engaged as they are, each in his own advancement, it is with greatest sacrifice that elevated spirits descend to earth. They husband their time with care, improving every moment; and only when compelled by love for their erring brothers do they hold intercourse with man.

The lower circles make no such sacrifice. They retain the strongest affection for the place of their birth, and linger around it with regret and dissatisfaction. Ever near, they are ready to communicate at all times, and, as the majority of communicators are derived from this source, a ready and complete explanation is furnished of the many low and ridiculous communications.

Such a view of spirit life should teach you not to place dependence on their words at the expense of your reason, which at all times should be your guide.

There is no hope for the suicide to better his condi tion by plunging recklessly into the unknown. Death should not be desired. It will meet you soon enough, at least before you are prepared. Your motto should be, I will leave the world better than I found it. Each should employ the talents he possesses. What use is death? If you should ask the disembodied suicide, who hung himself because mad with care or misfortune, if death had removed these, or bettered his situation, his answer would be a wail of agony. To avoid a few forebodings, he plunged into the sea of despair.

Men look for heaven at an indefinite period ahead, and think it forms no part of their daily lives. They do not endeavor to make heaven of their lives, each day becoming better. Heaven and hell are not localities, but conditions of mind. There is heaven on earth as well as the spheres; and earth is as well adapted for man's happiness as the spheres for the happiness of spirits. Heaven should begin on earth. In your minds you should nourish it, until it grows out into your life, modifying your actions, harmonizing your thoughts and desires. Some look far ahead, and then see a time when the weary soul will find rest. Discouraged and dissatisfied with the world, they can see no hope but in the future life. Then their fancy paints Eden, and they imagine that the good and great will meet to sing forever on golden harps the praise of God. Direful necessity compels them to see thus, for their clouded minds cannot perceive the grand scheme of causes which is at work elevating mankind. They see no good thing in the human heart. Disbelieving in the reforms of the day, they see only an end to toil and misery in the future life. Great wrongs exist which must be righted, and happiness eternal and supreme ushered into the world.

They labor under a great error, which has been ruinous in its results. Instead of bringing their religion right home, and embodying it in their lives, they have considered the world a vale of tears, to be endured until death sets the spirit free. Dives received his good things in this life; Lazarus his evil; and hence the more miserable they made themselves, the more they would enjoy after passing the "valley of the shadow of death."

Heaven is around you. The good deed, the kind word, the loving heart, — these are the elements of heaven. If you do a deed of charity, going, some cold morning, when the frosty mantle of snow lies deep and heavily — heavily as the heart of distress, — and from your store give to the needy, making comfortable the cheerless home, adding fuel to the fire, food to the empty larder; if you suppress the conflagration of passion; if you control the wayward desire which threatens to plunge you into immoral actions, — have you not created a heaven in your breast?

Do not look across the shade of death for heaven, for it is in your own hearts, and you can go on cultivating its presence, until your most secret thoughts will breathe nought but peace, love, and good will towards your fellow-men.

Here, too, is hell — ignorance, folly, and their resulting woe and wretchedness. Hell! Who coined that awful word? The combined crimes of mankind! Hell is the inharmonious relation to the laws of our being. It is for good, and not from revenge. Punishment is set along life's journey, as watch and police to guard our every footstep. We can go out of the path if we please; but the nettles sting, and the thorns tear us, until we gladly return. We see men every day just going far enough out to meet the jagged thorns. But better thus than not have the thorns, for just beyond yawn precipices of eternal destruction.

We will mention one relation more, and bring our present work to a close. Our relation we have not mentioned. The affections are transplanted, and thrive more unfettered with us than on earth. The ancients gave a happy illustration of the marriage relation by calling the two who entered into the union two halves, wandering around the world, at last meeting. But if the wrong halves united there was discord. The fable is true. If the right halves unite, conjugal love is perfectly gratified. It seeks no other object, but sees all that is perfect in its object, and never enjoys so much pleasure as in the society of its mate. Not that either loses existence. It is the perfect and eternal union of two individuals, without the least discordant jar. Such unions are rare on earth; but in the spheres all unions are such. We know that the loved one by our side is devoted to us for eternity; and our love is equally fervent. Together we share all blessings, and our joys are complete.

Patient reader, hoping to welcome you to this clime, of which, with feeble effort, we have endeavored to enlighten you, we bid you adieu.

THE END.

THE BANNER OF LIGHT:

A JOURNAL OF

ROMANCE, LITERATURE, AND GENERAL INTELLIGENCE,

AND ALSO

An Exponent of the Spiritual Philosophy of the Nineteenth Century.

PUBLISHED IN BOSTON, MASS.,

BY WILLIAM WHITE & CO.

LUTHER COLBY, Editor,

ASSISTED BY SOME OF THE ABLEST REFORMATORY WRITERS IN THE COUNTRY.

The distinctive features of the BANNER OF LIGHT are:

EDITORIALS — on subjects of general interest.

LITERARY DEPARTMENT. — In this Department Original Novellettes of reformatory tendencies will be given, and occasionally translations from the French and German.

ORIGINAL ESSAYS. — In this Department we shall publish from time to time Essays upon Philosophical, Scientific and Religious subjects.

REPORTS OF SPIRITUAL LECTURES, given by Trance and Normal Speakers.

MESSAGE DEPARTMENT. — Under this head we publisn weekly a variety of Spirit-Messages from the departed to their friends in earth-life, given through the instrumentality of MRS. J. H. CONANT, from the educated and the uneducated, which go to prove spiritual intercourse between the mundane and supermundane worlds.

All which features render the BANNER OF LIGHT a popular Family Paper, and at the same time the harbinger of a glorious Scientific Religion.

TERMS OF SUBSCRIPTION.

The BANNER OF LIGHT is published weekly, in quarto form (eight pages), on fine paper, for two dollars and fifty cents per annum, or one dollar twenty-five cents for six months, payable in advance.

Single copies can be purchased at all the principal Periodical Depots in the United States, at five cents per copy.

☞ Specimen copies sent free.

Address all Business Letters to "THE BANNER OF LIGHT."

WILLIAM WHITE & CO., Publishers,

158 Washington Street, Boston Mass.

VALUABLE BOOKS

ON

SPIRITUALISM,

PUBLISHED AND FOR SALE BY

WILLIAM WHITE & CO.,

AT THE

Banner of Light Office, No. 158 Washington Street,

BOSTON, MASS.

WORKS BY A. B. CHILD, M. D.

A B C of Life. 25 cents, postage 2c.
Soul Affinity. 15 cents, postage 2c.
The Bouquet of Spiritual Flowers, received chiefly through the mediumship of Mrs. J. S. Adams. $1.00, postage 16c.
Whatever Is, Is Right. $1.00, postage 16c.

WORKS BY A. J. DAVIS.

Answers to Ever-Recurring Questions from the People. (A Sequel to the Penetralia.) $1.25, postage 20c.

Free Thoughts concerning Religion: or NATURE VERSUS THEOLOGY. 15 cents, postage 2c.

Nature's Divine Revelations: A VOICE TO MANKIND. $2.50, postage 50c.

The Great Harmonia, in 5 volumes. Vol. 1, — THE PHYSICIAN. Vol. 2, — THE TEACHER. Vol. 3, — THE SEER. Vol. 4, — THE REFORMER. Vol. 5, — THE THINKER. $1.25 each, postage 20c. each.

The Harbinger of Health. $1.25, postage 20c.

The Harmonial Man: or, THOUGHTS FOR THE AGE. 30 cents, postage 6c. cloth 50 cents, postage 12c.

The History and Philosophy of Evil. In paper, 30 cents, postage 6c.; cloth 50 cents, postage, 12c.

The Magic Staff; an Autobiography of Andrew Jackson Davis. $1.25, postage 20c.

The Penetralia; being Harmonial Answers to Important Questions. $1.25, postage 24c.

The Philosophy of Special Providences: A VISION. 15 cents, postage 2c.

The Philosophy of Spiritual Intercourse; being an explanation of modern mysteries. 50 cents, postage 6c.; cloth, 75 cents, postage 12c.

WORKS BY DIFFERENT AUTHORS.

A Letter to the Chestnut Street Congregational Church, Chelsea, Mass., in Reply to its Charges of having become a reproach to the Cause of Truth, in consequence of a change of Religious Belief. By John S. Adams. 15 cents, postage 2c.

An Eye-Opener; or CATHOLICISM UNMASKED. By a Catholic Priest. 40 cts, postage free.

Answers to Charges of Belief in Modern Revelation, &c. By Mr. and Mrs. A. E. Newton. 10 cents, postage 2c.

Arcana of Nature; or the History and Laws of Creation. By Hudson Tuttle. 1st vol. $1.25, postage 16c.

Arcana of Nature; or the Philosophy of Spiritual Existence and of the Spirit World. By Hudson Tuttle. 2d vol. $1.00, postage 16c.

Blossoms of our Spring. A poetic work. By Hudson and Emma Tuttle. Price $1 00, postage 20c.

Familiar Spirits, AND SPIRITUAL MANIFESTATIONS: being a series of articles by Dr. Enoch Pond, Professor in the Bangor Theological Seminary, with a reply, by A. Bingham, Esq., of Boston. 15 cents, postage 4c.

Further Communications from the World of Spirits, on subjects highly important to the human family. By Joshua, Solomon and others. 50 cents, postage 8c.; cloth 75c., postage 12c.

Incidents in my Life. By D. D. Home, with an introduction by Judge Edmunds. $1 25, postage, free.

Legalized Prostitution; or Marriage as it Is, and Marriage as it should be, philosophically considered. By Chas. S. Woodruff, M. D. 75 cents, postage 16c.

Messages FROM THE SUPERIOR STATE. Communicated by John Murray, through J. M. Spear. 50 cents, postage 12c.

New Testament Miracles, and Modern Miracles. The comparative amount of evidence for each; the nature of both; testimony of a hundred witnesses. An Essay read before the Divinity School, Cambridge. By J. H. Fowler, 30 cents, postage 4c.

Plain Guide to Spiritualism. A Spiritual Hand-Book. By Uriah Clark. Cloth, $1.00, postage 16c.; paper, 75 cents, postage 8c.

Poems from the Inner Life. By Miss Lizzie Doten. Price, full gilt, $1.75, postage free; plain, $1.00, postage 16c.

Reply to the Rev. Dr. W. P. Lunt's Discourse against the Spiritual Philosophy. By Miss Elizabeth R. Torrey. 15 cents, postage 2c.

Spirit Manifestations: being an Exposition of Views respecting the Principal Facts, Causes and Peculiarities involved, together with interesting Phenomenal Statements and Communications. By Adin Ballou. Paper 50 cents, postage 6c.; cloth 75 cents, postage 12c.

The Bible; Is it of Divine Origin, Authority and Influence? By S. J. Finney. Cloth 40 cents, postage 8c.

The Fugitive Wife. By Warren Chase. 25 cents, postage free; cloth 40 cts. postage free.

The Gospel of Harmony. By Mrs. E. Goodrich Willard. Price 30 cents, postage 4c.

The History of Dungeon Rock. 25 cents, postage 2c.

The Kingdom of Heaven; OR THE GOLDEN AGE. By E. W. Loveland. 75 cents, postage 12c.

The "Ministry of Angels" Realized. A letter to the Edwards Congregational Church, Boston. By A. E. Newton. 15 cents, postage 2c.

The Philosophy of Creation; unfolding the Laws of the Progressive Development of Nature, and embracing the Philosophy of Man, Spirit, and the Spirit World. By Thomas Paine, through the hand of Horace Wood, medium. Cloth, 40 cents, postage 8c.

The Psalms of Life: a compilation of Psalms, Hymns, Chants, and Anthems, &c., embodying the Spiritual, Reformatory and Progressive sentiment of the present age. By John S. Adams. $1.00, postage 16c.

The Religion of Manhood; OR, THE AGE OF THOUGHT. By Dr. J. H. Robinson. Bound in muslin, 75 cents, postage 12c.

The Soul of Things; or PSYCHOMETRIC RESEARCHES AND DISCOVERIES. By William and Elizabeth M. F. Denton. Price, $1.25, postage 20c.

The Spirit Minstrel. A collection of Hymns and Music for the use of Spiritualists in their Circles and Public Meetings. Sixth edition, enlarged. By J. B. Packard and J. S. Loveland Paper, 25c., postage free; cloth 38 cents, postage free.

The Spiritual Reasoner. 50 cents, postage 12c.

Twelve Messages from the spirit of John Quincy Adams, through Joseph B. Stiles, medium, to Josiah Brigham. $1.50, postage 32c.

Who is God? A Few Thoughts on Nature and Nature's God, and Man's Relation thereto. By A. P. M'Combs. 10 cents, postage 2c.

Woodman's Three Lectures on Spiritualism, in reply to Wm. T. Dwight, D. D. 20 cents, postage 4c.

ENGLISH WORKS ON SPIRITUALISM.

Night Side of Nature: OR, GHOSTS AND GHOSTS-SEERS. By Catherine Crowe Price, $1.00, postage 20c.

Light in the Valley. MY EXPERIENCE IN SPIRITUALISM. By Mrs. Newton Crosland. Illustrated with about twenty plain and colored engravings. $1.00, postage 16c.

MISCELLANEOUS AND REFORM WORKS.

A False and True Revival of Religion. By Theodore Parker. 8 cents.

An Essay ON THE TRIAL BY JURY. By Lysander Spooner. Leather, $1.50, postage 20c.; cloth $1.00, postage 16c.; paper 75 cents, postage 8c.

A Sermon ON FALSE AND TRUE THEOLOGY. By Theodore Parker. 8 cents.

A Voice from the Prison, OR TRUTHS FOR THE MULTITUDE. By James A. Clay. 75 cents, postage 12c.

A Wreath for St. Crispin: being Sketches of Eminent Shoemakers. By J. Prince. 40 cents, postage 12c.

Battle Record OF THE AMERICAN REBELLION. By Horace E. Dresser, A. M. Price 25 cents, postage 2c.

Christ and the Pharisees upon the Sabbath. By a Student of Divinity. 20 cents, postage 4c.

Consumption. How to Prevent it, and How to cure it. By James C. Jackson, M. D. $2.00, postage 24c.

Eight Historical and Critical Lectures on the Bible. By John Prince. $1.00, postage 16c.

Eliza Woodson; OR, THE EARLY DAYS OF ONE OF THE WORLD'S WORKERS. A Story of American Life. Price, $1.25, postage free.

Eugene Becklard's Physiological Mysteries and Revelations. 25 cents, postage 4c.

Facts and Important Information for Young Men, on the subject of Masturbation. 12 cents, postage free.

Facts and Important Information for Young Women on the same subject. 12 cents, postage free.

Hesper, the Home Spirit. A Story of Household Labor and Love. By Miss Lizzie Doten. Price 80 cents, postage 12c.

Love and Mock Love. By Geo. Stearns. 25 cents, postage 4c.

Marriage and Parentage; or the Reproductive Element in Man, as a means to his Elevation and Happiness. By Henry C. Wright. $1.00, postage 20c.

Optimism, THE LESSON OF AGES. By Benjamin Blood. 60 cents, postage 12c.

Pathology of the Reproductive Organs. By Drs. Trall and Jackson. Price $4.00, postage 36c.

Peculiar. A Tale of the Great Transition. By Epes Sargent. Price $1.50, postage free.

Personal Memoir OF DANIEL DRAYTON. 25 cents. Cloth 40c.

Redeemer and Redeemed. By Rev. Charles Beecher. Price, $1.50, postage free.

Report OF AN EXTRAORDINARY CHURCH TRIAL: Conservatives *versus* Progressives. By Philo Hermes. 15 cents, postage 2c.

Six Years in a Georgia Prison. Narrative of Lewis W. Paine, who was the sufferer. Paper 25 cents, postage 4c.; cloth 40 cents, postage 12c.

www.ingramcontent.com/pod-product-compliance
Lightning Source LLC
LaVergne TN
LVHW010240110826
845151LV00004B/1346
* 9 7 8 1 4 2 5 5 2 4 0 7 4 *